M. Nakamura · P.M. Vanhoutte (Eds.)

Coronary Circulation in Physiological and Pathophysiological States

With 73 Illustrations, 1 in Color

Springer-Verlag
Tokyo Berlin Heidelberg
New York London Paris
Hong Kong Barcelona

Professor MOTOOMI NAKAMURA, M.D.
Research Institute of Angiocardiology and Cardiovascular Clinic,
Faculty of Medicine, Kyushu University, 3-1-1 Maidashi, Higashi-ku,
Fukuoka, 812 Japan

Professor PAUL M. VANHOUTTE, M.D.
Baylor College of Medicine, One Baylor Plaza, Texas Medical Center,
Houston, Texas 77030, U.S.A.

Library of Congress Cataloging-in-Publication Data
Coronary circulation in physiological and pathophysiological states / Motoomi
Nakamura, Paul M. Vanhoutte (eds.). p. cm. Contains the proceedings of
the International Symposium on the Coronary Circulation in Physiological and
Pathological States, held March 30, 1990 in Fukuoka, Japan, as a satellite symposium
to the 54th Annual Meeting of the Japanese Circulation Society.
ISBN-13: 978-4-431-68110-6   e-ISBN-13: 978-4-431-68108-3
DOI: 10.1007/ 978-4-431-68108-3
1. Coronary heart disease—Pathophysiology—Congresses. 2. Coronary circulation
—Congresses. I. Nakamura, Motoomi, 1927– II. Vanhoutte, Paul M. III. Inter-
national Symposium on the Coronary Circulation in Physiological and Pathological
States (1990: Fukuoka-shi, Japan) IV. Nihon Junkanki Gakkai. Gakujutus
Shūkai (54th: 1990: Fukuoka, Japan) [DNLM: 1. Coronary Circulation—congresses.
WG 300 C8217 1990] RC685.C6C634 1990 DNLM/NDLC for Library of Congress
90-10414

Typesetting: Best Set, Hong Kong

Preface

MOTOOMI NAKAMURA

As we approach the 21st century, ischemic heart disease is the major cause of death in most of the developed nations of the world. Since the 1970s, much effort and expense have led to designs of coronary thrombolytic therapy, percutaneous coronary angioplasty (PTCA), coronary artery bypass grafting, heart transplantation, automatic defibrillators, as well as to the formation of beta blockers and compounds which block the calcium channel. Socio-educational programs directed at exercise, diet, instruction in the risk factors of smoking, hyperlipidemia and hypertension have contributed to the decrease in the rate of morbidity and mortality of patients with ischemic heart disease. However, the first clinical event of ischemic heart disease, the so-called "heart attack" and sudden cardiac death continues to present problems, as the mechanisms involved in these events are poorly understood.

It has long been thought that ischemic heart disease is the sequence of an organic fixed atherosclerotic obstruction of the epicardial coronary arteries and the role of coronary vasomotion has been given much less attention. Recent clinical and laboratory animal studies revealed that increased tonus and spasm of the large epicardial coronary arteries are the cause of various stages of ischemic heart disease. The role of coronary vasospasm in the development of unstable angina, sudden cardiac death and acute myocardial infarction remains open to debate.

Pharmacophysiological studies showed that the epicardial large coronary artery contributes only 5% to regulation of normal coronary flow. From laboratories of the pharmaceutical industry have come various antianginal drugs which do increase coronary flow in normal animals, but these same coronary dilators are not always effective in patients with angina pectoris. Thus, an animal model suitable for screening antianginal drugs is not available.

Two volumes on Coronary Circulation In Health and Disease by Dr. Donald Gregg in 1950 and by Dr. Melvin L. Marcus in 1983, prove to be major reference books which have contributed greatly to a better understanding of the pathophysiology of coronary circulation. More recent acquisition of evidence for endothelium-dependent relaxing and contracting factors has greatly elucidated ischemia-related factors. Cell culture techniques, molecular cell biology studies, and an animal model of coronary spasm and atherosclerosis have greatly contributed to our knowledge of the pathophysiology of ischemia-related heart disease.

Looking back over my 35 years as a cardiologist, in the laboratory and at the bedside, it became apparent that while much of the enthusiasm and efforts directed toward treatment of heart patients were related to events following the onset of clinical symptoms, little or less attention was placed on a search for factors that initiate and sustain the so-called "heart attack." Examination of the pathophysiology of ischemic heart disease must be made both in animal and clinical studies, in a complementary manner.

Described in the present volume are the most recent advances and achievements in studies on 1) autonomic regulation of the coronary circulation in conscious animals, 2) characteristics and possible origins of blood velocity waveforms of epicardial and intramyocardial coronary circulation, 3) effects of low density lipoproteins on endothelial modulation of coronary tone, 4) characteristics of vascular smooth muscle cell membranes and their modifying factors, 5) endothelium, platelets, and vascular occlusion, 6) defective arterial flow reserve in atheroscleroses, 7) an animal model of coronary spasm and the progress of coronary atherosclerosis, 8) effects of atherosclerosis and the regression on regulation of vascular vasomotion, and 9) biological and molecular biological aspects of angiogenesis in coronary collateral development.

Distinguished investigators from the United States, West Germany, and Japan participated in an International Symposium on Coronary Circulation in Physiological and Pathological States, held on March 30, 1990 in Fukuoka, Japan, in honor of my retirement from Kyushu University. Their lectures are included in this volume.

Ongoing research includes, 1) the design of techniques to prevent a restenosis after PTCA, 2) noninvasive modalities to assess abnormalities of microcirculation in the human heart, 3) intra- and intercellular events regulating growth of the coronary atheroma, vascular tone, angiogenesis, and expression of vasoactive genes, 4) development of gene therapy to alter cell functions in cases of pathology of the coronary circulation, 5) seeding of transformed cells onto artificial grafts, 6) development of animal models, including transgenic ones, and 7) role of central nerve functions in regulating coronary circulation.

I do hope this volume will be found useful for the education and training of young physicians and graduate students who must understand the basic principles underlying mechanisms of coronary heart disease and be aware of the recent advancements in the field of coronary circulation in health and disease. Practitioners will also find it of value for direct application to clinical problems. A sound, comprehensive training for researchers and clinical cardiologists is most important if one is to go on to carry out innovative and meaningful studies. A researcher needs oceans of patience during long pauses with little or no apparent advance in the research. They must continue to apply themselves until their objective has been reached. Studies on cell and molecular biology plus those on the whole animal are needed when attempting to better comprehend vascular systems. It is the cardiologist who needs to look closely at coronary atherosclerosis, a major factor in ischemic heart disease.

It is with deep regret that the late Dr. Melvin L. Marcus, who was Professor of Medicine at the University of Iowa, could not be with us for the Symposium. My dear and respected friend passed away on October 19, 1989. He had been asked to honor us by being chief editor of this volume.

I express my sincere and heartfelt gratitude for the contributions of the speakers who made both the symposium and this volume possible. Finally, I am greatly obliged to all my colleagues, coworkers, technical and secretarial staff, and most of all to my dear family for their kind support and understanding over all these years.

Preface

PAUL M. VANHOUTTE

This monograph represents the Proceedings of an International Symposium on "The Coronary Circulation in Physiological and Pathological States," which was held on March 30, 1990, in Fukuoka, Japan, as a satellite symposium to the 54th Annual Meeting of the Japanese Circulation Society. The symposium brought together scientists from all over the world and from all over Japan. They had two things in common: their fascination with the complexity of coronary circulation in health and disease, and their admiration for Professor Nakamura, the co-editor of this book. Indeed, the symposium was held in the honor of Professor Nakamura, at the time of his retirement as Head of the Institute of Angiocardiology and Cardiovascular Clinic of the School of Medicine of Kyushu University in Fukuoka. The participants wanted to express their respect to a man who, throughout a brilliant and prestigious career, contributed more than anybody else to the understanding of coronary vasospasm and the role that it plays in cardiac diseases. They did so during a most stimulating scientific session. Hence their contribution constitutes an up to date overview of the major events and factors affecting coronary circulation and will be of interest to all scholars, whether clinicians or researchers, who follow Professor Nakamura in the quest for the understanding of the causes of coronary vasospasm and heart disease.

I would like to thank Professor Nakamura for allowing me not only to contribute to the symposium in his honor, but also to help him edit this monograph. As an editor, I would like to thank all the authors for their impressive and timely contributions, as well as acknowledging the collaboration of Springer-Verlag Tokyo for the highly professional handling of the manuscripts. As a participant of the symposium, I would like to thank Dr. Akira Takeshita and the organizing committee of the symposium most sincerely for their gracious hospitality.

Contents

XII   Contents

List of Contributors

The page numbers refer to the page on which the contribution begins

CHAPTER 1

Characteristics and Possible Origins of Blood Velocity Waveforms of the Epicardial and Intramyocardial Coronary Circulation in the Ventricles and the Atria

F. Kajiya[1]

Summary. We measured blood velocities in small epicardial coronary arteries and veins of the ventricles and the left atrium, and intramyocardial arteries and veins using our optical-fiber laser Doppler velocimeter which provides excellent access to vessels. The phase opposition of velocity waveforms between coronary arteries and veins was consistent for both left and right ventricles when velocity measurements were performed in a small artery just before its penetration into myocardium, and in a small vein just after its emergence. The phase opposition was more marked in intramyocardial vessels. Diastolic displacement of blood from superficial veins to deeper portions was frequently observed. Atrial contraction caused a transient sharp decrease in arterial flow of the left atrial coronary arteries (systolic dip), and a prominent systolic flow in atrial veins. Thus, the effect of muscle contraction and relaxation on coronary arterial and venous flows may be fundamentally similar in the left and right ventricles, and in the left atrium. The phase opposition indicates the importance of intramyocardial capacitance vessels as a determinant of phasic coronary arterial and venous flows. To investigate the functional characteristics of the intramyocardial capacitance vessels, we analyzed the change in venous flow following changes in coronary arterial inflow. It was shown that during diastole the intramyocardial capacitance vessels have two functional components, unstressed volume and ordinary capacitance. Unstressed volume is defined as the volume of blood in a vessel at zero transmural pressure, and it was approximately 5% of the volume of the myocardium. When the unstressed volume was saturated, the coronary inflow was decreased significantly, compared with that for the unsaturated condition. The systolic coronary venous outflow showed a significant, positive correlation with the total displaceable blood volume stored in the intramyocardial capacitance vessels. Thus, the increase in intramyocardial blood volume decreases the coronary artery inflow, whereas it enhances coronary venous outflow.

Introduction

The phasic flow in the left coronary artery is diastolic-predominant. Scaramucci (1689, cited by Porter [1]) hypothesized that the deeper coronary vessels are

---

[1] Department of Medical Engineering and Systems Cardiology, Kawasaki Medical School, 577 Matsushima, Kurashiki, Okayama, 701-01 Japan

compressed by the contraction of the surrounding muscle fibers with displacement of intramyocardial blood into the coronary veins. These intramyocardial vessels refill from the aorta during diastole. To prove this hypothesis, it is necessary to investigate the coronary arterial inflow and venous outflow of the myocardium. Since Anrep et al. [2] investigated circulation of the coronary arteries and veins more than sixty years ago by making blood flow measurements using a hot wire method, there have been few reports [3–5] describing simultaneously the arterial inflow and venous outflow. This is partly because until recently, measurement of coronary venous flow using conventional methods including electromagnetic flowmeter has been difficult, and also because the vein was regarded as the only conduit of coronary venous outflow. Direct measurement of intramyocardial blood velocity and pressure has been hampered by difficulty in finding an appropriate route of access of a transducer into myocardial vessels of a beating heart. However, by employing new techniques, several investigators have succeeded in measuring blood velocity, vascular diameter, and blood pressure in the microcirculation of the superficial myocardium, i.e., the direct visualization technique [6–9] and the servo-nulling micromanometer [10,11].

The laser Doppler velocimeter (LDV) with an optical fiber is a powerful new tool for measurement of both coronary arterial and venous flow [12–14]. The most important advantage of the LDV method over conventional velocimeters is ease of access to vessels, even easily collapsible veins and vessels located in deeper parts of myocardium. In this paper, we describe the optical arrangement of LDV with an optical fiber, and the routes of access of the fiber probe to various coronary vessels. We report the following results: (1) blood velocity waveforms in small epicardial coronary arteries and veins of the ventricles and left atrium, and in intramyocardial arteries and veins, (2) the functional characteristics of intramyocardial capacitance vessels, and (3) the effect of blood pooling in intramyocardial capacitance vessels on coronary arterial inflow and venous outflow.

Principle of Fiber Optic Laser Doppler Velocimeter

Figure 1 shows our newly revised system of LDV. The He-Ne laser beem (632.8 nm, 5 mW) is divided by a beam splitter (BS). The greater part of the initial light passed by the BS is focused onto the entrance of a graded-index multimode fiber (crad diameter: 125 or 62.5 μm, core diameter: 50 μm), and transmitted through the fiber into the blood stream. Part of the light scattered back by flowing erythrocytes is collected by the same fiber, and transmitted back to the entrance. The other part of the initial light divided at the BS is used as a reference beam. Two frequency shifters (82 and 78 MHz) are interposed on the path of the incidence and reference beams to differentiate the forward flow from the reverse. Thus, the difference of the shifter frequencies (82 − 78 = 4 MHz) indicates zero flow velocity. When a Doppler-shift frequency is greater than 4 MHz the direction of blood flow is toward the fiber tip, and when it is less than 4 MHz the direction is away from the fiber tip. The optical heterodyning is made

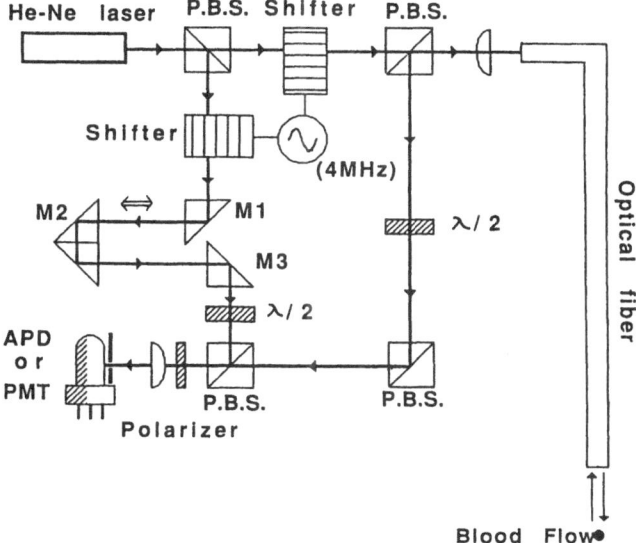

Fig. 1. Schematic diagram of the laser Doppler velocimeter with optical fiber (revised type). PBS, polarization beam splitter; M1,M2,M3, mirrors; APD, avalanche photodiode; PMT, photomultiplier

by mixing the Doppler-shift signal from the moving erythrocytes with the reference beam. The photocurrent from the photodetector (APD) is fed into a spectrum analyzer to detect the Doppler-shift frequency.

The back-scattered light signal has a Doppler-shift frequency Δf from 4 MHz given by

$$\Delta f = 2nV\cos\theta/\lambda$$

where V is the blood flow velocity; n is the refractive index of blood, approximately 1.33; θ is the angle between the fiber axis and the axis of the blood stream, and λ is the laser wavelength of 632.8 nm in a free space. When θ is 60°, the Doppler-shift frequency of 1 MHz corresponds approximately to a blood velocity of 48 cm/s. The sample volume of our system is approximately $\pi \times 0.05^2 \times 0.1$ mm^3, and the temporal resolution is 8 ms [15–17].

Three Different Routes of Access of the Fiber Probe to Coronary Vessels

Our laser Doppler velocimeter provides excellent access to coronary vessels of the moving heart. We used three different routes of access of the fiber probe according to the measuring objectives (Fig. 2), i.e., epicardial large coronary vessels, epicardial small artery and vein, and intramyocardial artery and vein.

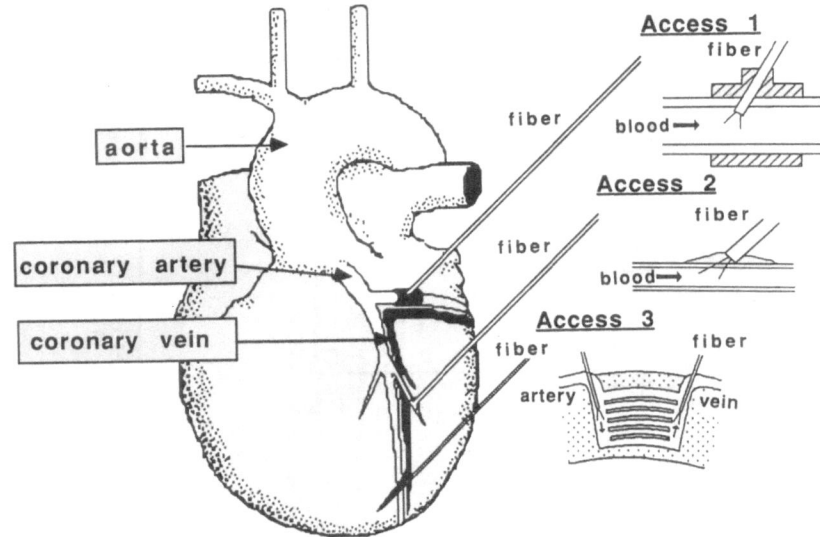

Fig. 2. Three routes of access of the fiber probe to blood vessels. Route 1 for blood velocity measurements in large and middle-sized epicardial coronary arteries and veins. Route 2 for blood velocity measurements in small epicardial coronary arteries and veins. Route 3 for blood velocity measurements in intramyocardial arteries and veins

For measurement of blood flow velocity in large and middle-sized epicardial coronary arteries and veins, we placed a cuff around the vessel and inserted the fiber into the vessel through a small hole of the cuff (route 1) [18]. The fiber tip was moved stepwise across the vessel to obtain the velocity profile across the vessel. For measurement of blood flow velocity in small epicardial arteries and veins whose walls were thin enough to be transparent to laser light, we placed the fiber tip on the outer surface of the vessel and fixed it by a drop of cyanoacrylate (route 2) [19]. For measurement of blood flow velocity in intramyocardial arteries and veins, we inserted the fiber into the vascular lumen from a position just penetrating into the myocardium, and introduced the fiber into a deeper portion (route 3) [19].

Blood Velocity Waveforms in Small Epicardial Coronary Arteries and Veins of the Ventricles and Left Atrium, and in Intramyocardial Vessels

Blood Velocity Waveforms in Small Epicardial Coronary Arteries and Veins of the Left Ventricle

A number of reports are available which describe phasic flow characteristics in the epicardial coronary arteries. It is well-known that left coronary arterial and venous flows are dramatically affected by cardiac contraction. However, there have been few reports on the characteristics of phasic coronary arterial inflow and venous outflow of the myocardium. For the investigation of phasic myocardial perfusion, it is necessary to evaluate phasic blood flow velocities in the peripheral coronary artery and vein, since proximal coronary arterial and large venous flows may differ from the peripheral flows due to the capacitance effect of epicardial arteries and veins. Chilian and Marcus [20] assessed the phasic pattern of coronary blood velocity in the distal coronary artery by high-frequency ultrasound pulsed Doppler velocimeter. They characterized the velocity waveform in distal coronary arteries in two ways: (1) the amount of flow in systole is decreased by more than half, and the amount in diastole is increased by 20% compared with the flow in the proximal coronary artery, and (2) the midsystolic flow is retrograde. Hellenbrand et al. [21] measured intravascular venous flow and found that the velocity waveform was influenced by the autonomic nervous system. However, there are no data available for a small vein at the site just after emergence from the myocardium.

We measured phasic flow velocity in small epicardial arteries at a position just before they penetrated the myocardium, and the velocity in small veins just after they emerged from the myocardium, using access route 2. Figure 3 shows an example of the velocity waveforms in a small epicardial artery and vein of the left ventricle. As indicated by Chilian and Marcus [20], the velocity waveform in the peripheral epicardial artery was almost exclusively diastolic, and reverse flow was frequently observed during the early half of systole. However, the small epicardial veins of the left ventricle exhibited a systolic-predominant flow which is the general characteristic of coronary venous flow. Compared with the great

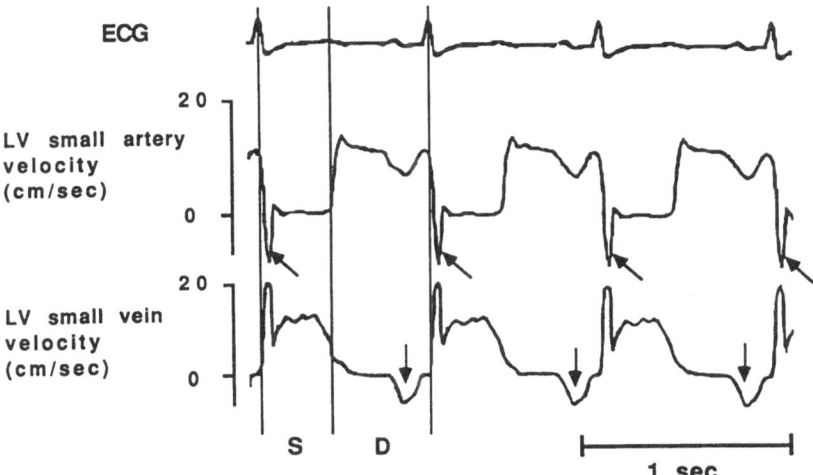

Fig. 3. An example of velocity waveforms in a small epicardial artery of the left ventricle just before its penetration into the myocardium, and in a small vein just after its emergence from the myocardium. The *arrows* indicate reverse flow velocities in the artery and vein

cardiac vein (GCV) or coronary sinus flow, the onset of the small coronary venous flow was earlier, the flow acceleration was higher, and the diastolic flow component was much smaller and frequently reversed. Therefore, the difference in phase of the velocity waveforms was more prominent between the artery and the vein peripherally than in the proximal portions of large epicardial arteries and veins.

Blood Velocity Waveforms of Small Arteries and Veins of the Right Ventricle

Although it is well established that the systolic flow component is greater in the right coronary artery (RCA) than in the left coronary artery (LCA) [22,23], the phasic patterns of the arterial inflow and venous outflow of the right ventricular myocardium are unclear because of the lack of such data. To solve this problem, we measured, in the dog, the blood flow velocity in the distal right coronary artery just before its penetration into the myocardium, as well as that in a small coronary vein just after its emergence from the myocardium, using access route 2 [24] (Fig. 2).

Figure 4 shows a representative tracing of blood velocities in an epicardial, peripheral small artery and vein. The systolic flow component in the peripheral right coronary arteries was much smaller than that observed in their proximal portions, indicating a significant contribution of epicardial compliance to blood velocity waveforms. The systolic flow component of the peripheral artery of the right ventricle was even larger, when it was compared with that of the left ventricle.

The blood flow velocity waveform in the epicardial small veins of the right ventricle was always characterized by the systolic-predominant pattern. As

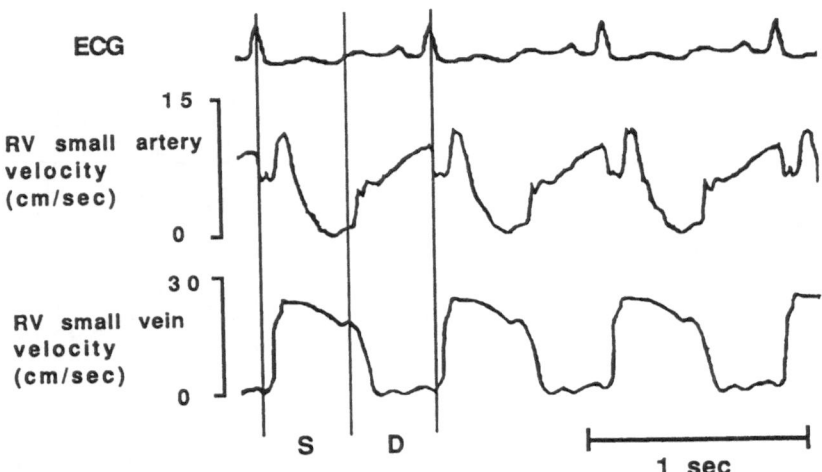

Fig. 4. A representative tracing of blood velocities in a small artery of the right ventricle just before its penetration into myocardium, and in a small vein just after emergence from the myocardium. RV, right ventricle

shown in Fig. 4, blood velocity increased with rise in right ventricular pressure, and decreased with right ventricular relaxation. This venous systolic flow wave may be caused by the displacement of blood from intramyocardial capacitance vessels into the vein, by contraction of the right ventricular myocardium.

We conclude that the velocity waveforms in peripheral small artery and vein of the right ventricle are fundamentally diastolic- and systolic-predominant, respectively, like their velocity waveforms for the left ventricle. Thus, the impeding effect on arterial flow, and the expressing effect on venous flow by myocardial contraction of the right ventricle seem more powerful than we have inferred.

Velocity Waveforms in Atrial Small Arteries and Veins

In order to obtain some insight into the nature of the mechanical force acting on the intramyocardial vascular beds, we measured the flow velocity waveforms in small arteries and veins by access route 2 [25]. In particular, we intended to examine the influence of atrial contraction on velocity waveforms of atrial artery and vein. The left atrial appendage of the dog was displaced gently to expose small branches of the artery and/or vein, in order to obtain access for the optical fiber probe.

Figure 5 shows a representative tracing of blood flow velocities in left atrial small artery and vein under control conditions. The atrial arterial velocity increased with aortic pressure, and gradually decreased after reaching a peak during midsystole. The blood flow velocity waveform showed a dicrotic notch, as seen in the aortic pressure waveform. Thus, the outline of blood velocity waveform resembled the pattern of aortic pressure during ventricular ejection. However, a sharp transient decrease in inflow velocity, that is a dip corresponding to the atrial contraction phase, was always present during late ventricular diastole. The dip appeared with the rise in left atrial pressure, and ceased at the end of the left ventricular isovolumic contraction. The atrial venous blood velocity rapidly increased with atrial contraction, and decreased with atrial relaxation. After a midsystolic nadir, blood flow velocity gradually increased prior to the onset of the atrial contraction. The major forward velocity wave of the vein showed a reciprocal relation to the dip of the arterial velocity waveform.

The main observations were: (1) the velocity waveform of the arterial flow resembled the aortic pressure pattern, except during the atrial contraction, (2) atrial contraction caused a transient sharp decrease in arterial flow velocity, and (3) the velocity waveform of the atrial veins was characterized by a prominent atrial systolic velocity wave. We thus conclude that atrial myocardial contraction impedes arterial inflow (the systolic dip), and promotes venous outflow from the atrial capacitance vessels.

Velocity Waveform in Intramyocardial Arteries and Veins

In addition to their function as conduits, epicardial coronary arteries function as capacitors for blood flow. Because of possible difference in hemodynamics

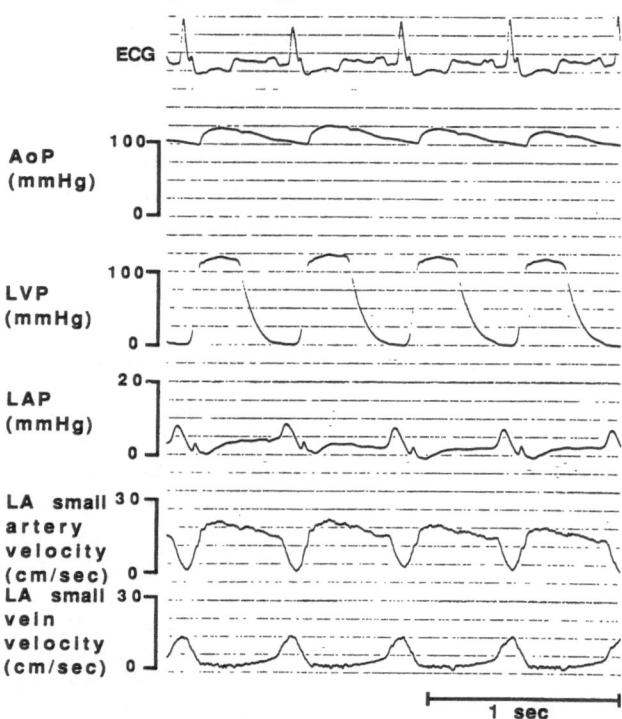

Fig. 5. A representative recording of blood flow velocities in left atrial (*LA*) small artery and vein. AoP, aortic pressure; LVP, left ventricular pressure; LAP, left atrial pressure. (From [25] with permission of the American Heart Association)

between epimyocardium and endomyocardium, the flow in epicardial coronary vessels does not provide direct information on intramyocardial flow dynamics. Therefore, measurements of intramyocardial blood flow are needed for better understanding of coronary circulatory physiology. A trans-illumination method is a powerful tool to evaluate superficial epimyocardial hemodynamics. However, measurements of phasic blood velocities in the deeper myocardial vessels have been limited by methodology. Recently, we measured the blood velocity in the intramyocardial small arteries and veins, and in deep sites of the septal artery, using access route 3, shown in Fig. 2 [26–28].

Figure 6 shows an example of the velocity patterns in the septal artery and in an intramyocardial small vein of the dog. Measurements were obtained during administration of isoproterenol to enhance the effect of myocardial compression on velocity waveforms. In this case, the depth of the measured position of the septal artery was 18 mm from the orifice of the septal artery, and that of the intramyocardial vein was approximately 2 mm beneath the cardiac surface. The most prominent characteristic of the septal arterial velocity waveform was systolic reverse flow, i.e., reverse flow was divided into two components,

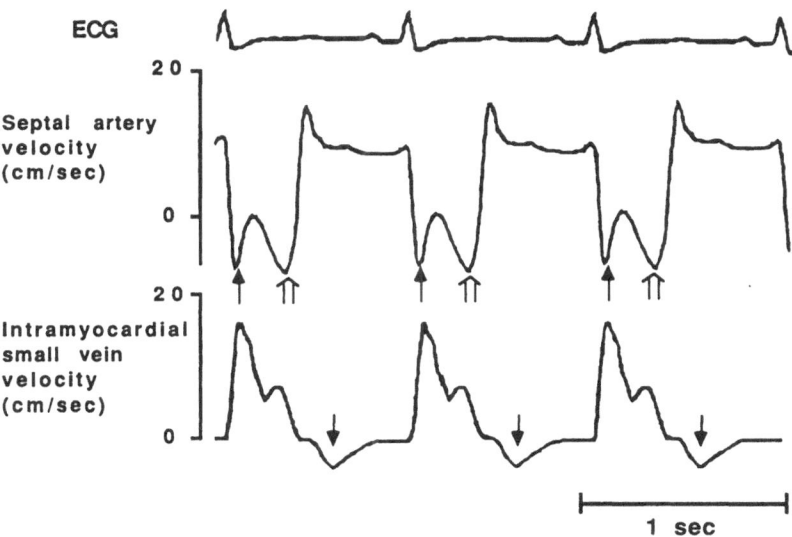

Fig. 6. An example of velocity patterns in the septal artery (18 mm depth from its orifice), and in a small intramyocardial vein during administration of isoproterenol

isovolumic and mid-and/or late-systolic reverse flows. The systolic forward flow wave of venous flow was divided into two components. It was concluded that systolic forward flow has a reciprocal relation with reverse flow in intramyocardial arteries. During diastole, there may be suction of blood from superficial veins into the deeper portions, since diastolic reverse flow was frequently observed in intramyocardial coronary veins as well as in epicardial small veins just after their emergence from the myocardium (Fig. 4). Figure 7 illustrates possible blood flow directions during one cardiac cycle, based on earlier reports [8,20,29] and on our own observations [27,28]. There may be translocation of blood between epimyocardium and endomyocardium.

Concluding Remarks

We observed the following: (1) The phase opposition of velocity waveforms between coronary arteries and veins was consistent for both left and right ventricles when velocity measurements were performed in a small artery just before its penetration of the myocardium, and in a small vein just after its emergence. The ratio of systolic to diastolic velocity components was higher in the right than in the left ventricle. (2) Atrial contraction caused a transient sharp decrease in arterial flow velocity of small coronary arteries of the atria (the systolic dip), and a prominent atrial systolic flow in small coronary veins of the atria. (3) The phase opposition between arterial and venous flows was more remarkable in intramyocardial vessels. The systolic reverse flow in the artery showed a reciprocal

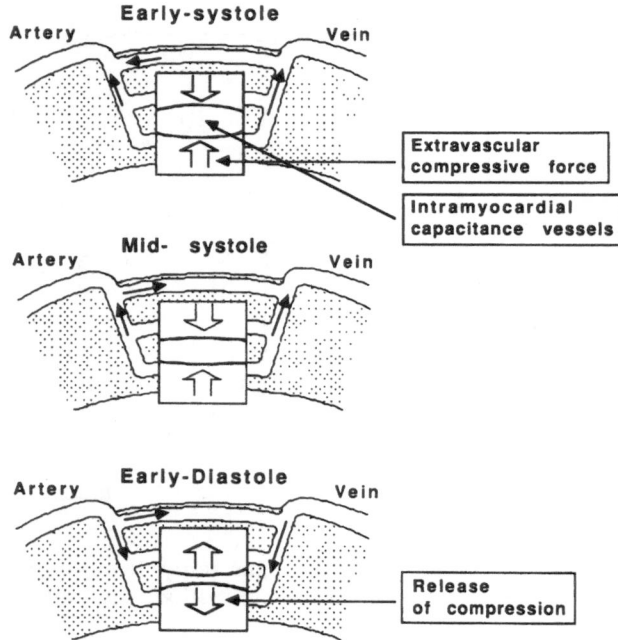

Fig. 7. A schematic drawing of possible blood flow directions in intramyocardial arteries and veins during one cardiac cycle as postulated by our observations and earlier reports

relation to the systolic forward flow in the vein. Suction of blood from superficial veins to deeper portions may occur during diastole.

The effect of muscle contraction and relaxation on coronary arteries and veins may be fundamentally similar in the left and right ventricles, and the left atrium.

Intramyocardial Capacitance Vessels, and Their Effect on Coronary Arterial Inflow and Venous Outflow

The phase opposition between coronary arterial and venous flows indicates that the blood which flows into the intramyocardial capacitance vessels during diastole is expelled into the coronary veins during systole. Thus, the intramyocardial capacitance vessels are considered to be an important determinant of both the phasic coronary arterial and venous flows [30].

In order to investigate the functional characteristics of the intramyocardial capacitance vessels, we analyzed the response of the great cardiac vein (GCV) and the interventricular venous flows following changes in coronary arterial inflow. We used a similar protocol to analyze the effect of intramyocardial capacitance vessels on coronary arterial inflow and venous outflow.

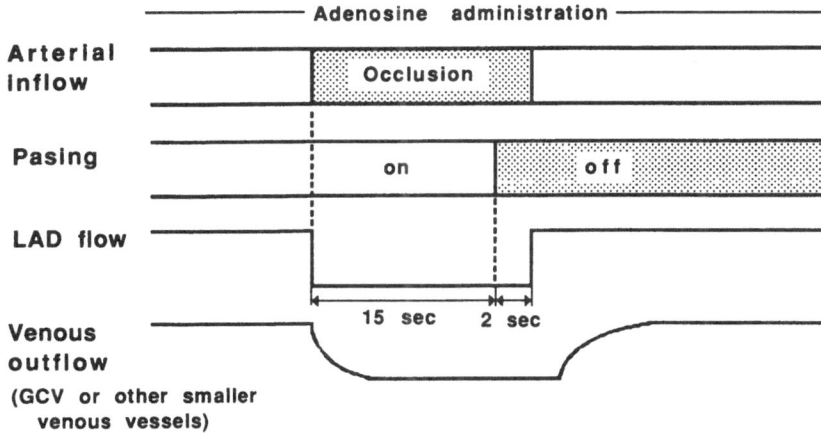

Fig. 8. Experimental protocol. Coronary arterial inflow was shut off, and coronary venous flow decreased. Fifteen seconds after arterial inflow occlusion, a long diastole was induced. Two seconds later, arterial perfusion was resumed, and the perfusion pressure was increased in a stepwise fashion: (1) for the analysis of functional characteristics of intramyocardial vessels, the response of coronary venous flow to stepwise increases in arterial perfusion pressure was observed; (2) to investigate the effect of capacitance vessels on arterial inflow, the time course of arterial flow after reperfusion was analyzed; (3) for the analysis of the effect of capacitance vessels on venous outflow, the decaying process of coronary venous flow was used. (Redrawn from [25] with permission of the American Heart Association)

Functional Properties of Intramyocardial Capacitance Vessels

In order to analyze the characteristics of the intramyocardial capacitance vessels, the blood flow velocity of the GCV was measured after a stepwise increase in the coronary arterial pressure during prolonged diastole [31].

In anesthetized, AV-blocked dogs, thoracotomy was performed, the peripheral portion of the GCV was isolated, and the optical fiber tip was inserted into the vessel by access route 1 in Fig. 2. Both the left main coronary artery (LM) and the left anterior descending coronary artery (LAD) were cannulated and connected to a reservoir to regulate perfusion pressure. In our earlier study, only the LAD was cannulated and connected to the reservoir. Coronary inflow and perfusion pressures were measured at the peripheral portion of the cannula inserted into the LAD. The experimental procedure is shown in Fig. 8. During continuous infusion of adenosine into the coronary artery, the cannulae were occluded to shut off LAD flow. The blood velocity in the GCV decreased and reached a minimal steady value within 15 s. A long diastole was induced by cessation of pacing. The cannulae were reopened 2 s after the cessation of pacing, and the perfusion pressure was increased stepwise to a preset target pressure. The time course of GCV flow velocity was analyzed after the initiation of reperfusion.

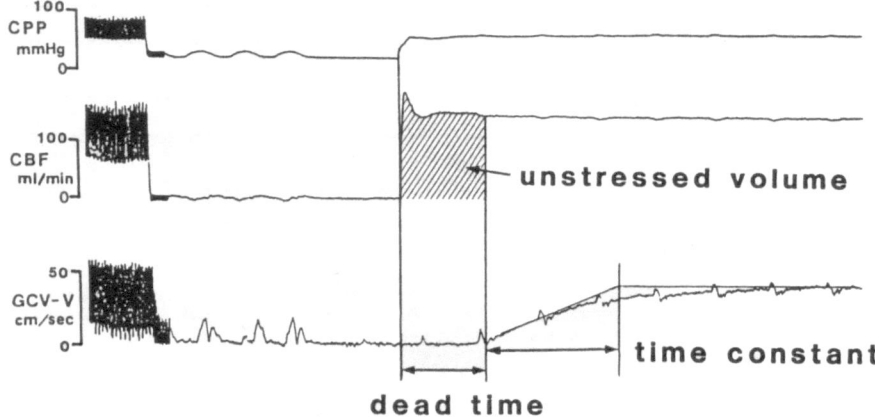

Fig. 9. A representative tracing of variables recorded during a trial. The coronary venous flow was absent for approximately 1 s (dead time), and increased with a first order time delay after reperfusion. CPP, coronary perfusion pressure; CBF, coronary arterial blood flow in left anterior descending coronary artery (LAD); GCV-V, great cardiac venous velocity. (Redrawn from [25] with permission of the American Heart Association)

The time course of the coronary hemodynamic data is displayed in Fig. 9. After occlusion of the coronary inflow, the blood velocity in the GCV decreased and reached a steady minimum value. With the cessation of pacing it fell to zero. After reopening the inflow, it remained absent for a few seconds (dead time). It then reappeared, and increased with a first-order time delay. The presence of dead time indicates the existence of an unstressed volume in the intramyocardial vascular compartments, which is defined as the volume of blood in a vessel at zero transmural pressure. The time constant of the first-order delay relates to the product of resistance and capacitance of the diastolic coronary circulation with minimal vasomotor tone.

The mechanical lumped model illustrated in Fig. 10 was adopted as the optimal , simple model for explaining the results of these animal experiments. The model consists of a combination of the unstressed volume, the resistances, and the capacitance. The unstressed volume was estimated from the coronary arterial inflow during the dead time, and averaged 5.2 ml/100g LV. The value of capacitance was obtained by dividing the time constant by the resistance. The mean value of capacitance was 0.08 ml/mmHg per 100 g LV. With vessels embedded in tissue as intramyocardial vessels, transmural pressure at a volume less than the unstressed volume may be negative. To return the vessels from a collapsed or a semi-collapsed to a cylindrical form requires only a small change in intraluminal pressure. This characteristic of the vessels may contribute to the unstressed volume. This volume is distributed continuously in the coronary circulation system, but not homogeneously. The unstressed volume may be mostly distributed in veins and capillaries because of their distensibility. Therefore, some inflow into the unstressed volume may both refill the intramyocardial

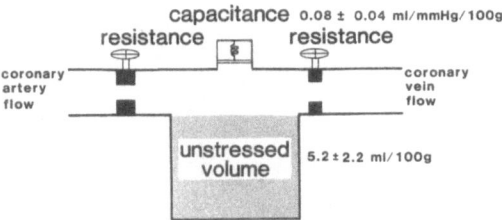

Fig. 10. Mechanical lumped model representing the characteristics of the intramyocardial capacitance vessels and estimated values of the unstressed volume and the capacitance. (Redrawn from [25] with permission of the American Heart Association)

capacitance vessels and allow nutrient exchange. The unstressed volume is approximately 5% of the myocardium, and the time constant in relation to ordinary capacitance is approximately 1 s, although they are both pressure-dependent.

Effect of Intramyocardial Capacitance on Coronary Arterial Inflow

To evaluate the effect of intramyocardial capacitance vessels on coronary arterial inflow, the last part of the protocol in Fig. 8 was used [19,32,33]. The pressure-flow relation was analyzed under different conditions in the intramyocardial capacitance vessels, i.e., unstressed volume-filled and unstressed volume-unfilled conditions. When the unstressed volume is filled, coronary venous flow occurs, but when it is unfilled, coronary venous flow does not occur. Thus, the blood-storage condition of unstressed volume, filled or unfilled, is indicated by the presence or absence of coronary venous flow.

The target pressure after reopening the perfusion cannulae was set at seven different levels. Figure 11 shows a representative example of the seven time courses of the LAD pressure, and the flow responses resulting from changes in the target pressure. Two time courses of GCV flow corresponding to high and lower target pressures are also included in the same figure. Each GCV flow had a dead time prior to flow resumption after reperfusion, but the length of this dead time depended on the target pressure; i.e., the higher the target pressure, the shorter the dead time. As indicated in Fig. 11, two time points were determined to analyze the pressure-flow relationships under conditions of unfilled and filled-unstressed volume. Figure 12 shows the mean results of pressure-flow relationships under different blood-storage conditions of unstressed volume. The pressure-flow relationships show excellent linearities and correlation coefficients in the 7 dogs, ranging from 0.97 to 0.99. The zero-flow pressure intercepts for the unstressed volume-unfilled and filled phase were 19.9 ± 2.2 and 18.9 ± 1.8 mmHg, respectively. There was no statistical difference. The inverse of the slope for the unstressed volume-filled phase was 0.69 ± 0.08 mmHg·ml^{-1}·min^{-1}, which was significantly higher than that for the unfilled phase, 0.55 ± 0.66 mmHg·ml^{-1}·min^{-1} (P < 0.01). These results indicate that

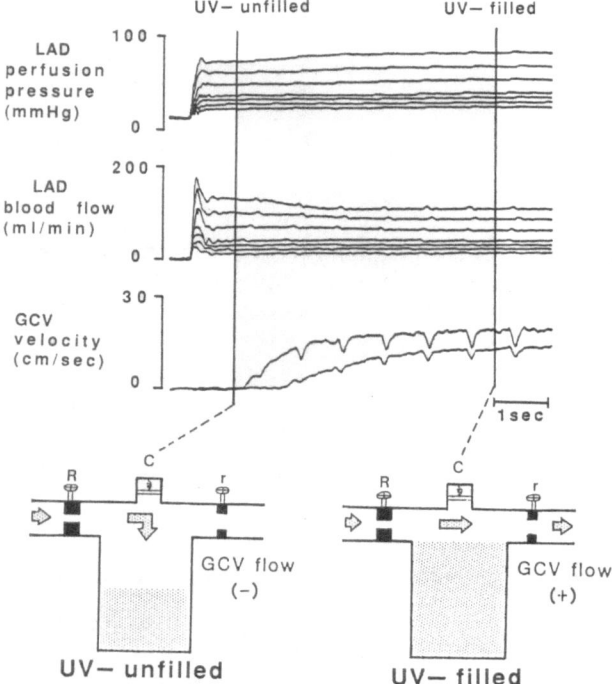

Fig. 11. A representative example of 7 time courses of pressure and flow responses in left anterior descending coronary artery (*LAD*), caused by changing target pressures. Two time courses of great cardiac venous (*GCV*) flows corresponding to higher and lower target pressures are also displayed. Two points for analysis of the pressure-flow relationships under different blood-storage conditions of unstressed volume are indicated by vertical lines with their conceptual explanation models (below). UV, unstressed volume; LAD, left anterior descending coronary artery; GCV, great cardiac vein. (Redrawn from [32] with permission of the American Physiological Society)

the filling condition of intramyocardial capacitance vessels plays an important role in diastolic coronary arterial inflow. An increase in the blood volume above the unstressed volume in intramyocardial capacitance vessels has an impeding effect on diastolic coronary arterial inflow.

Effect of Intramyocardial Capacitance on Coronary Venous Outflow

To investigate the effect of intramyocardial capacitance vessels on coronary venous outflow, the initial part of the protocol shown in Fig. 8 was used [19,32,33]. The time course of GCV flow velocity was analyzed after occlusion of the cannulae. In order to determine the relationship between the GCV velocities before occlusion and those during the expression of blood after occlusion, prior to occlusion, GCV velocities were changed to various values by altering the perfusion pressure.

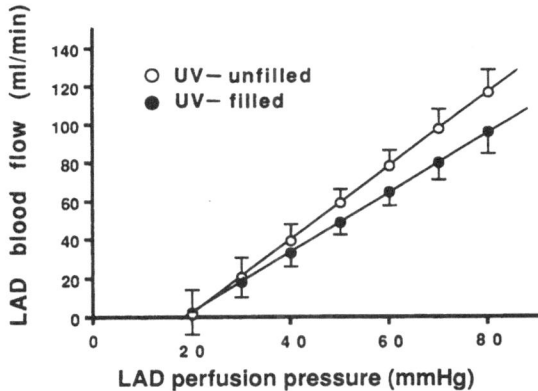

Fig. 12. Averaged results of pressure-flow relationships for different blood storage conditions. *Vertical bars*, 95% confidence limits for seven different pressures; UV, unstressed volume; LAD, left anterior descending coronary artery. (From [32] with permission of the American Physiological Society)

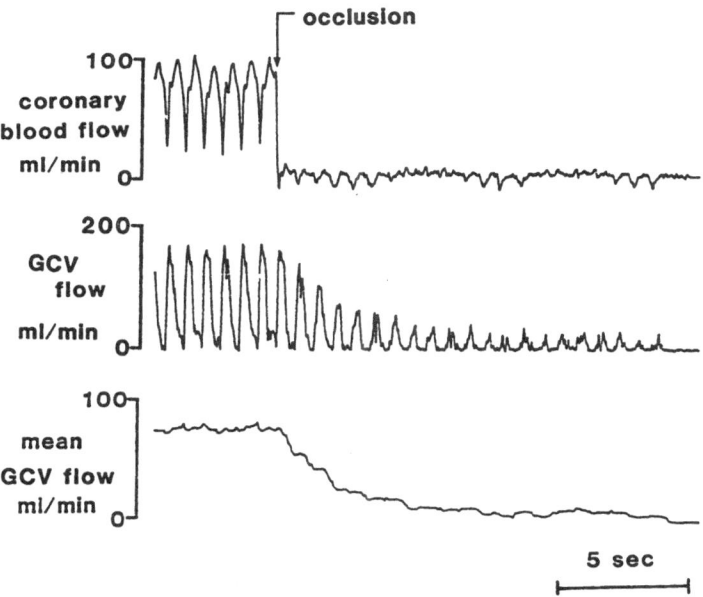

Fig. 13. A typical example of the decaying process of the coronary venous flow velocity after coronary inflow occlusion. The great cardiac venous (*GCV*) flow decreased exponentially. We integrated the GCV flow after occlusion of arterial inflow. (From [19] with permission of the Japanese Circulation Society)

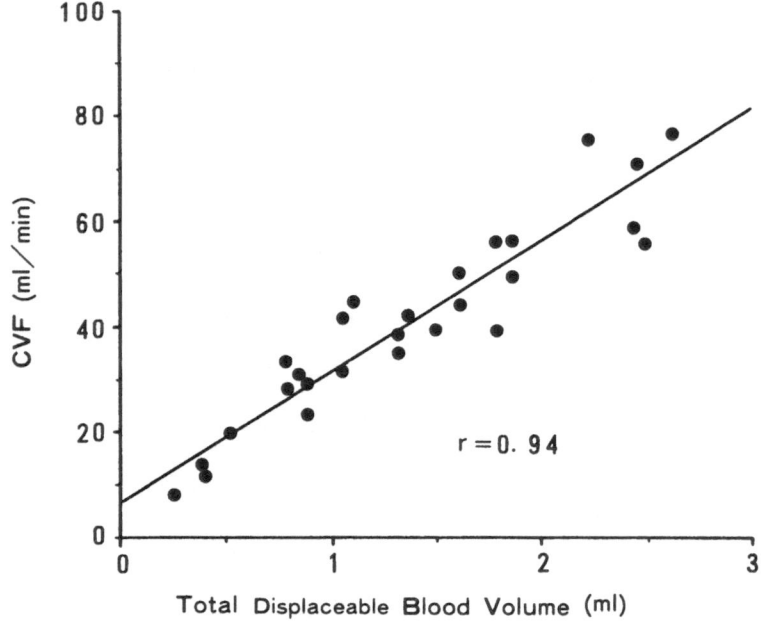

Fig. 14. The relationships between great cardiac vein flow velocities and total displaceable blood volume in intramyocardial vessels. The relationship showed a significant correlation. (From [19] with permission of the Japanese Circulation Society)

Figure 13 shows the time course of decay of GCV flow velocity after coronary inflow occlusion. We integrated GCV flow after the occlusion of cannula. The value of this integration provides the total displaceable volume stored in the intramyocardial capacitance vessels prior to occlusion. Fig. 14 shows the relationship between the total displaceable volume and the coronary venous flow prior to occlusion. The correlation coefficient was significantly high (p < 0.01). This indicated that the coronary venous outflow is closely related to the total displaceable blood volume stored in the intramyocardial vessels, prior to ventricular contraction.

Concluding Remarks

(1) During diastole the intramyocardial capacitance vessels have two functional components, unstressed volume and ordinary capacitance.
(2) When the unstressed volume was saturated, the diastolic coronary inflow was significantly decreased, compared with that for the unsaturated condition.
(3) Systolic coronary venous outflow is closely related to the total displaceable blood volume stored in intramyocardial capacitance vessels. Therefore, the increase in intramyocardial blood volume decreases the coronary arterial inflow, whereas it enhances coronary venous outflow. This is an important mechanical control of coronary circulation.

Acknowledgments. The author wishes to acknowledge the collaboration of Katsuhiko Tsujioka, Yasuo Ogasawara, Osamu Hiramatsu, Masami Goto, Akihiro Kimura, Tokunori Yamamoto, Shinichiro Tadaoka, Masanobu Nakai, Yoshifumi Wada, Xiao-Liang Chen, Swee C. Tjin, Keiichiro Mito, Go Tomonaga, and Noritake Hoki, and also Mayumi Yokomizo for secretarial assistence. This work was partly supported by Grant-in-Aid 01480253 for scientific research from the Ministry of Education, Science, and Culture, Japan, and a research grant for cardiovascular diseases (1A-1) from the Ministry of Health and Welfare.

References

1. Porter WT (1898) The influence of the heart-beat on the flow of blood through the walls of the heart. Am J Physiol 1: 145–163
2. Anrep GV, Cruickshank EWH, Downing AC, Sabba RA (1927) The coronary circulation in relation to the cardiac cycle. Heart 14: 111–133
3. Chilian WM, Marcus ML (1984) Coronary venous outflow persists after cessation of coronary arterial inflow. Am J Physiol 247: H984–H990
4. Spaan JAE (1982) Intramyocardial compliance studies by venous outflow at arterial occlusion (abstract). Circulation 66: II–42
5. Kajiya F, Hiramatsu O, Mito K, Tadaoka S, Ogasawara Y, Tsujioka K (1990) Evaluation of coronary blood flow by fiber-optic laser Doppler velocimeter. In: Kajiya F, Klassen GA, Spaan JAE, Hoffman JIE (eds) Coronary circulation. Springer, Tokyo, pp 43–54
6. Tillmanns H, Ikeda S, Hansen H, Sarma JSM, Fauvel J-M, Bing RJ (1974) Microcirculation in the ventricle of the dog and turtle. Circ Res 34: 561–569
7. Tillmanns H, Steinhausen M, Leinberger H, Thederan H, Kubler W (1981) Pressure measurements in the terminal vascular bed of the epimyocardium of rats and cats. Circ Res 49: 1202–1211
8. Ashikawa K, Kanatsuka H, Suzuki T, Takishima T (1986) Phasic blood flow velocity pattern in epimyocardial microvessels in the beating canine left ventricle. Circ Res 59: 704–711
9. Kanatsuka H, Lamping KG, Eastham CL, Dellsperger KC, Marcus ML (1989) Comparison of the effects of increased myocardial oxygen consumption and adenosine on the coronary microvascular resistance. Circ Res 65: 1296–1305
10. Nellis SH, Liedtke AJ, Whitesell L (1981) Small coronary vessel pressure and diameter in an intact beating rabbit heart using fixed-position and free-motion techniques. Circ Res 49: 342–353
11. Chilian WM, Eastham CL, Marcus ML (1986) Microvascular distribution of coronary vascular resistance in beating left ventricle. Am J Physiol 251: H779–H788
12. Tanaka T, Benedek GB (1975) Measurement of the velocity of blood flow (in vivo) using a fiber optic catheter and optical mixing spectroscopy. Appl Optics 14: 189–196
13. Kajiya F, Hoki N, Tomonaga G, Nishihara H (1981) A laser-Doppler-velocimeter using an optical fiber and its application to local velocity measurement in the coronary artery. Experientia 37: 1171–1173
14. Kilpatric D, Linderer T, Sievers RE, Tyberg JV (1982) Measurement of coronary sinus blood flow by fiber-optic laser Doppler anemometry. Am J Physiol 242: H1111–H1114

15. Kajiya F, Mito K, Ogasawara Y, Tsujioka K, Tomonaga G (1984) Laser Doppler blood flow velocimeter with an optical fiber and its applications to detailed measurements of the coronary blood flow velocities. Proc SPIE 494: 25–31
16. Kajiya F, Hiramatsu O, Mito K, Ogasawara Y, Tsujioka K (1987) An optical-fiber laser Doppler velocimeter and its application to measurements of coronary blood flow velocities. Med Prog Technol 12: 77–85
17. Kilpatrick D, Kajiya F, Ogasawara Y (1988) Fibre optic laser Doppler measurement of intravascular velocity. Australas Phys Eng Sci Med 11: 5–14
18. Kajiya F. Tomonaga G, Tsujioka K, Ogasawara Y, Nishihara H (1985) Evaluation of local blood flow velocity in proximal and distal coronary arteries by laser Doppler method. J Biomech Eng 107: 10–15
19. Kajiya F, Tsujioka K, Ogasawara Y, Mito K, Hiramatsu O, Goto M, Wada Y, Matsuoka S (1989) Mechanical control of coronary artery inflow and vein outflow. Jpn Circ J 53: 431–439
20. Chilian WM, Marcus ML (1985) Effects of coronary and extravascular pressure on intramyocardial and epicardial blood velocity. Am J Physiol 248: H 170–H178
21. Hellenbrand WK, Klassen GA, Armour JA, Sezerman O, Paton B (1986) Autonomic nervous system regulation of epicardial coronary vein systolic and diastolic blood velocity as measured by a laser Doppler velocimeter. Can J Physiol Pharmacol 64: 1463–1472
22. Gregg DE, Khouri EM, Rayford CR (1965) Systematic and coronary energetics in the resting unanesthetized dog. Circ Res 16: 102–113
23. Lowensohn HS, Khouri EM, Gregg DE, Pyle RL, Patterson RE (1976) Phasic right coronary artery blood flow in conscious dogs with normal and elevated right ventricular pressures. Circ Res 39: 760–766
24. Hiramatsu O, Wada Y, Yamamoto T, Yanaka M, Kimura A, Ogasawara Y, Tsujioka K, Kajiya F (1989) Similar phasic characteristics of artery inflow into and vein outflow from myocardium between left and right ventricles (abstract). Circulation 80: II–549
25. Kajiya F, Tsujioka K, Ogasawara Y, Hiramatsu O, Wada Y, Goto M, Yanaka M (1989) Analysis of the characteristics of the flow velocity waveforms in left atrial small arteries and veins in the dog. Circ Res 65: 1172–1181
26. Mito K, Ogasawara Y, Hiramatsu O, Wada Y, Goto M, Tadaoka S, Tsujioka K, Kajiya F (1987) Evaluation of velocity waveform in an intramyocardial small artery and vein by laser Doppler method (abstract). Circulation 76: IV–386
27. Mito K, Ogasawara Y, Hiramatsu O, Wada Y, Tsujioka K, Kajiya F (1988) Evaluation of blood flow velocity waveforms in intramyocardial artery and vein by laser Doppler velocimeter with an optical fiber. In: Manabe H, Zweifach BW, Messmer K (eds) Microcirculation in circulatory disorders. Springer, Tokyo, pp 525–528
28. Hiramatsu O, Mito K, Kajiya, F (1990) Evaluation of the velocity waveform in intramyocardial small vessels. In: Kajiya F, Klassen GA, Spaan JAE, Hoffman JIE (eds) Coronary circulation. Springer, Tokyo, pp 169–172
29. Carew TE, Covell JW (1976) Effect of intramyocardial pressure on the phasic flow in the intraventricular septal artery. Cardiovasc Res 10: 56–64
30. Spaan JAE (1985) Coronary diastolic pressure-flow relation and zero flow pressure explained on the basis of intramyocardial compliance. Circ Res 56: 293–309
31. Kajiya F, Tsujioka K, Goto M, Wada Y, Chen X-L, Nakai M, Tadaoka S, Hiramatsu O, Ogasawara Y, Mito K, Tomonaga G (1986) Functional characteristics of intramyocardial capacitance vessels during diastole in the dog. Circ Res 58: 476–485

32. Goto M, Tsujioka K, Ogasawara Y, Wada Y, Tadaoka S, Hiramatsu O, Yanaka M, Kajiya F (1990) Effect of blood filling in intramyocardial vessels on coronary arterial inflow. Am J Physiol 258: H1042–H1048
33. Tsujioka K, Goto M, Hiramatsu O, Wada Y, Ogasawara Y, Kajiya F (1990) Functional characteristics, stics of intramyocardial capacitance vessels and their effects on coronary arterial inflow and venous outflow. In: Kajiya F, Klassen GA, Spaan JAE, Hoffman JIE (eds) Coronary circulation. Springer, Tokyo, pp 89–97

CHAPTER 2

Biological and Molecular Biological Aspects of Angiogenesis in Coronary Collateral Development

W. Schaper[1]

Summary. We have studied over several years the development of coronary collateral vessels in pigs and dogs following progressive coronary artery occlusion by ameroid-, hydraulic-and (intermittent) pneumatic occlusion. Preexistent collateral vessels respond with growth by DNA-synthesis, mitosis and proliferation of endothelium and smooth muscle cells. Experiments with chronically-implanted, remote-controlled, hydraulic-occluders in canine hearts showed that when the time between onset of stenosis and complete occlusion was 3 days, mostly endothelial cells entered the S-phase of the reproductive cell cycle. When it took 5 days, mostly smooth muscle cells were undergoing DNA-synthesis. In the dog heart, this growth occurs preferentially in epicardial vessels. Preexistent arterioles and small arteries are completely remodeled during the growth process, and the internal elastic lamina is digested first (probably by invading leucocytes), to allow expansion of the growing vessel. After the expansion phase is over, a new elastic lamina is formed. Monocyte invasion is a typical occurrence in growing collaterals of dog. Monocytes are known to produce a host of growth factors that may stimulate endothelium and smooth muscle to divide. The chemotactic stimulus for monocytes to adhere to endothelium may be ammonia from the deamination of endothelium. Adenosine is a metabolite of ATP which is broken down in ischemia. Monocyte involvement is less well observed in pig collaterals. In the pig heart, numerous small vessels, located throughout the entire LV-wall thickness with a slight preference for the subendocardium, respond to progressive coronary occlusion and ischemia. The entire microvasculature of the risk region in the pig heart responds with enlargement by growth. This is a useful response because of the paucity of the preexistent collateral network in the pig heart: when a critical degree of stenosis is reached, the minimum capillary resistance falls, thereby ensuring nearly adequate tissue perfusion for only a small increase in collateral flow.

Monocytes were detected in dog collaterals by direct inspection (scanning electron microscopy of opened whole mounted collaterals), a method not applicable to the much smaller pig vessels. Monoclonal antibodies against human monocytes did not cross-react with pig monocytes.

Dog and pig hearts also differ after the completion of vascular growth. After chronic left circumflex occlusion the peripheral coronary pressure equals the pressure at the

---

[1] Max-Planck-Institute, Department of Experimental Cardiology, D-6350 Bad Nauheim, Federal Republic of Germany

origin of the collateral (= stem pressure = $0.8 \times$ aortic pressure). Although this is also true for the pig, i.e. PCP = Pstem, stem pressure is much lower, indicating that pig collaterals originate from much smaller vessels.

Mature dog collaterals can increase their initial tissue mass by a factor of 50.

Another approach to the study of collateral growth is intermittent coronary occlusion as introduced by Franklin and Tomoike. We occluded the left circumflex coronary artery in the pig (using a chronically-implanted pneumatic-occluder) for 2 min every 30 min, using the Tomoike-scheme. More than 400 occlusions were necessary to cause a significant increase in collateral flow, and to avoid systolic bulging (2D and M-mode echo). Unfortunately, a control group receiving only the implant achieved the same result with only two test occlusions after 4 weeks. Extracardiac collaterals (adhesions) complicate this experiment in pigs.

Growth by DNA-synthesis and mitosis usually occurs following stimulation by peptide-mitogens. We have searched for a heart-derived growth factor, and we have isolated two heparin-binding growth factors from normal pig, canine and bovine heart. These have a very high sequence homology with beta-ECGF which is involved in the growth of collaterals; the stored tissue hormone is released by proteolysis of the basement membrane, and new beta-ECGF is produced by upregulation of its gene. In situ hybridization with a heterologous beta-ECGF cRNA-probe showed upregulated gene expression only in *growing* pig blood vessels.

We are currently searching for a myocardial ischemia-derived myocyte factor that upregulates genes for peptide growth factors. The search strategy is based not on knowledge on the protein level, but rather on expected differences in transcription of genes in the collateralized versus non-collateralized parts of the heart, using a subtractive cloning approach.

The successful identification of new and known peptide growth factors and competence factors necessitates studies designed to unravel their interaction. It is not presently known how the six factors that are known to be angiogenic interact in the heart following the onset of ischemia. Preliminary data suggest an interaction between growth factors and coagulation factors. Stimulated endothelial cells produce urokinase and tissue plasminogen activator, which pave the way for new endothelial cells. Growth factors like TGF-beta, PDGF and PDECGF can also be carried into the potentially ischemic region via platelets which adhere at damaged collateral endothelium, and are the richest known source of these factors.

Introduction

In many patients the natural history of coronary artery disease is significantly influenced by the presence and development of coronary collaterals. Patients suffering from chronic stable angina pectoris often have occluded major coronary arteries and well-developed collaterals that can meet the basal needs of the heart for blood flow and nutrients, but not extra demands. Occlusion of the main left coronary artery without contractile dysfunction at rest due to the existence and development of collaterals has been described. It is well known that slowly progressive coronary artery disease favors the development of collaterals, but acute thrombotic occlusion does not. However, if a high grade stenosis has been present for some time prior to thrombotic occlusion, some protection is offered by collaterals that enlarge in response to stenosis.

The old anatomists and pathologists knew that collaterals respond to myocardial ischemia, especially in reversible ischemia, and Hunter is credited with the observation that collaterals go when and where they are needed.

We and others [1–3] have shown that collaterals do not enlarge passively by stretching, but actively by growth, i.e., DNA replication and mitosis of endothelial and smooth muscle cells. We have shown [4] that collateral development stops when the cardiac myocyte is no longer ischemic, however, this occurs before the normal coronary reserve is reached, i.e., at maximal coronary vasodilation, collateral flow is significantly less than coronary flow to a normal coronary region.

The detailed molecular and cell-biological mechanism in collateral vessel growth are unknown, but a few interesting pieces of a very complex puzzle are known.

Vascular Growth Factors Are Present in the Heart

We have shown that heparin-binding-growth-factors (HBGF), i.e. hormones that induce vessel growth in avascular organs (cornea), or increase vascularization (chorio-allantoic membrane) are also present in normal bovine and porcine hearts [5]. We have extracted these hormones [acidic (beta ECGF) and basic fibroblast growth factor from the HBGF family] and confirmed their biological activity in an endothelium-mitogen test, and their structural fidelity by amino acid sequencing. The existence of these potent growth factors in normal hearts opens several unresolved questions:

1. Under normal physiological conditions all cell populations in the heart are quiescent, i.e., they do not divide [6]. A barrier must therefore be assumed to exist between the potent mitogens and their receptors. Another explanation could be that the receptors are down-regulated under physiological conditions.
2. Although some observations point to a storage of HBGFs in the extracellular matrix, the molecular structure of both HBGFs makes it unlikely that these hormones can leave the cell because they lack a signal peptide, i.e., a sequence that functions to escort the peptide through the cell membrane during secretion. A few other growth factors and cytokines share this structural feature, e.g., interleukin 1 and platelet derived endothelial growth factor [7]. It has been hypothesized that the HBGFs are co-transported with other molecules across the cell membrane, or that dying cells release the hormone that is then stored in the matrix. It also remains unexplained that the protein is found in normal myocardium, but no transcript is present, i.e., the genes coding for the HBGFs are switched off.

Ischemia Up-Regulates Genes Coding for Growth Factors

Progressive experimental coronary occlusion in the porcine heart leads to increased transcription of mRNA for acidic FGF but only in the endothelium of growing microvessels (W. Schaper and R. Kandof 1990, unpublished data). This is an interesting observation which clarifies the fact that the ischemic cardiac myocyte produces other tissue factors that are able to up-regulate the expression of the acidic FGF-gene in microvessels. This factor is probably a competence factor that enables the endothelial cell to start an autocrine loop; upon stimulation by the ischemia-related mitogen-activator it produces acidic FGF that triggers mitosis. It is possible that the newly produced a-FGF need not leave the cell because it may meet its receptor inside the cell while on its way to the outer membrane.

Transforming Growth Factor β is Up-Regulated in Ischemic Myocardium

TGF-β causes neovascularization in previously avascular tissue, such as cornea. On a cellular level, TGF-β may act as a mitogen or as a mitogen inhibitor [8]. It plays a role in differentiation, and it is a potent attractant of monocytes and macrophages, that in turn are able to produce growth factors. In pigs with ameroid-induced coronary occlusion, we found increased expression on the protein (Western blot) and on the transcription level (Northern blot) [9]. In situ hybridization showed the transcript in cardiac myocytes. Although the signals were clearly separated from background, not all animals presumed to represent similar developmental stages showed evidence of increased expression. We assume that the time window for TGF-β in the process of collateralization is narrow.

The Role of Ischemia in Recalling a Fetal Pattern of Growth Factor Gene Expression

On angiographic images, collaterals often stand out because of their tortuous cork-screw-like pattern, i.e., they are much longer than necessary. Histological examination shows that the tortuosity is related to a subintimal layer of longitudinal smooth muscle. Although all arterial walls are built with a helical layer of subintimal muscle, the helices of normal arteries deviate only slightly from the usual circumferential arrangement. In collaterals, the "helices" are almost perpendicular to the circumference. We interpret this finding that ischemia recalls the embryonic pattern of vascular development, where both radial and longitudinal growth occurs at the same time. The increase in vessel length cannot be accommodated in the adult non-growing heart, hence the cork-screw-like appearance. The molecular patterns of genes expressing regulation of embryonal vascular growth are insufficiently known at present. Members of the HBGF-family seem to play an important role [10].

A Strategy to Isolate Unknown Growth Factors

The classical way of searching for biologically active peptides and proteins is to grind large amounts of relevant tissue, and find extremely minute amounts.

Biological assays often use up most of the precious material, and it takes a long time before the amino acid sequence can be determined. Gene structure is deduced from the amino acid sequence, and oligonucleotide probes have been constructed to search for the gene in a cDNA-library. After obtaining a complete (or sufficiently long) cDNA, studies on the regulation of gene expression can be carried out. A more powerful approach often used today is called "reversed genetics" because the gene is known before the protein is isolated. The process starts with isolation of mRNA from relevant tissue, followed by reverse transcription. The cDNA so obtained is hybridized against mRNA from a different source. In the case of regional ischemia, the cDNA of the ischemic region is hybridized against mRNA of a normal region of the same heart, and only single stranded DNA that is unique for the ischemic region is further cloned. We use this approach presently in our lab, to study collateralization and vascular development.

Collateral Development is not Possible Without Tissue Remodeling

Stimulation of endothelium by growth factors and subsequent mitosis increases the tissue mass. The affected small arteriole is a relatively firm structure built to withstand arterial pressures, and the new cells are not easily accommodated. The new cells will in fact reduce the size of the vascular lumen if the arterial structure is not actively remodeled by controlled destruction. Years ago [1] when we first described the histological development of collaterals, we were impressed with the amount of damage found in these vessels, and we stated that collaterals developed while repairing injury that we believed was caused by the relatively long lasting stretch of ischemic vasodilatation. However, other types of long-lasting vasodilation did not lead to injury, and did not result in much larger vessels (W. Schaper 1990, unpublished data). What we had described was the phase of active remodeling.

Remodeling proceeds in two phases:

(1) controlled destruction of the arterial structure to accommodate new endothelial and smooth muscle cells, creating a much larger vessel, and
(2) restructuring of the arterial bed to accommodate the larger artery. Most of the developing small arteries are surrounded by connective tissue and cardiac myocytes. At least two thirds of the circumference of epicardial collaterals is in close contact with cardiac myocytes, and since they impede the radial expansion they are actively destroyed and finally removed. These myocytes undergo a non-ischemic degeneration that differs from ischemic necrosis in that the mitochondria are not acutely damaged, and that the contractile proteins disappear first as if their expression was down-regulated. Those myocytes that are not completely destroyed finally become much smaller.

The molecular and cell-biological mechanisms responsible for remodeling are not completely known. The induction of vascular growth, whether in wound healing or tumor vascularization, involves the production of tissue plasminogen activator that leads to the formation of plasmin, a protease believed to play an important role in tissue remodeling. Plasminogen activators are produced by stimulated endothelium, and they probably aid digestion of the extracellular matrix that glues smooth muscle cells together. Elastase produced by invading leucocytes is needed to cut the internal elastic lamina, and macrophages produce collagenase that digests the perivascular collagen. Sometimes growth is so rapid that some smooth muscle cells become necrotic. Their removal by phagocytosis creates the necessary space to accommodate new cells.

Experimental Models in Relation to the Clinical Situation

In the human heart, two types of collateral circulation exist in the presence of coronary artery disease: a small number of large epicardial vessels, and numerous small endomural/subendocardial vessels. Fulton [11] has described a subendocardial plexus of collateral connections that occasionally show larger vessels. In many patients collaterals had developed in atrial muscle. These are often relatively large in caliber, and quite tortuous.

In experimental animals also, all classes of size and localization are found, but each species tends to favour either large and few epicardial vessels, or small and numerous endomural vessels. In the canine heart (the best known) most of the collaterals that develop in response to slowly progressing left circumflex coronary occlusion, are visible on the epicardial surface. The canine heart has specialized for the "large-and-few" type of collateral circulation. Visible only with X-rays, are connections between the septal artery and the right posterior descendents of the LCS, and between the right coronary artery and the LCX. These connections do not belong to the "numerous-small" type, but do not reach the caliber of the epicardial vessels either.

Porcine and equine hearts have specialized for the "numerous-small" type. In experimental LCX-occlusion in the pig heart, sometimes a few larger, traceable collaterals can be seen by naked eye when injected, and often by "soft" X-rays, on the endocardial surface of the posterior papillary muscle [12].

Much of this does not apply to microvessels, which react with growth in the entire risk region. However, since the reaction of so many vessels is only a small percentage of their normal size, remodeling is much less dramatic than in canine epicardial collaterals with an increase in diameter by a factor of 20.

The guinea pig is an interesting species which is so well endowed with collaterals that coronary occlusion does not lead to infarction [13]. Flow measurement in isolated guinea pig hearts showed that occlusion of a major coronary artery (right interventricularis anterior) did not even lead to a reduction of tissue perfusion.

Conclusion

No experimental animal exhibits the broad human repertoire of reactive vascular development following slowly progressing coronary stenosis. Dogs can be used to study the epicardial large vessel response, and pigs can be used to study the endomural numerous small vessel response. If manipulation of the collateral response should appear feasible in the future, both species should be studied. The study of vasculogenesis in the guinea pig appears to be of value.

References

1. Schaper W (1971) The Collateral Circulation of the Heart. North-Holland Publishing, Amsterdam
2. Schaper W (ed) (1979) The Pathophysiology of Myocardial Perfusion. Elsevier/North-Holland Biomedical, Amsterdam
3. Schaper W, Görge G, Winkler B, Schaper Jutta (1988) The collateral circulation of the heart. Prog Cardiovasc Dis 31: 57–77
4. Schaper W, Flameng W, Winkler B, Wüsten B, Türschmann W, Neugebauer G, Carl M, Pasyk S (1976) Quantification of collateral resistance in acute and chronic experimental coronary occlusion in the dog. Circ Res 39: 371–377
5. Quinkler W, Maasberg M, Bernotat-Danielowski S, Lüthe N, Sharma HS, Schaper W (1989) Isolation of heparin binding growth factors from bovine, porcine, and canine hearts. Eur J Biochem 181: 67–73
6. Schaper W, De Brabander M, Lewi P (1971) DNA-synthesis and mitoses in coronary collateral vessels of the dog. Circ Res 28: 671–679
7. Usuki K, Heldin NE, Miyazono K, Ishikawa F Takaku F, Westermark B, Heldin CH (1989) Production of platelet-derived endothelial cell growth factor by normal and transformed human cells in culture. Proc Natl Acad Sci USA 86: 7427–7431
8. Roberts AB, Sporn MB, Assoian RK, Smith JM, Roche NS, Wakefield LD, Heine VI, Liotta LG, Falanga V, Kehre JH, Fauci AS (1986) Transforming growth factor type beta: rapid induction of fibrosis and angiogenesis in vivo and stimulation of collagen formation in vitro. Proc Natl Acad Sci USA 83: 4167–4171
9. Wünsch M, Sharma HS, Bernotat-Danielowski Sabine, Schott RJ, Schaper Jutta, Bleese N, Schaper W (1989) Expression of transforming growth factor beta-1 (TGF-β1) in collateralized swine heart (abstract). Circulation 80 (Suppl): 1802
10. Risau W, Ekblom P (1986) Production of heparin-binding angiogenesis factor by the embryonic kidney. J Cell Biol 103, 1101–1107
11. Fulton WFM (1965) The Coronary Arteries, Thomas Publisher, Springfield
12. Schaper W, Vandesteene R (1967) The rate of growth of interarterial anastomoses in chronic coronary artery occlusion. Life Sci 6: 1673
13. Schaper W (1984) Experimental infarcts and the microcirculation. In: Hearse DJ, Yellon DM (eds) Therapeutic Approaches to Myocardial Infarct Size Limitation. Raven, New York, pp 79–90

Large Coronary Artery Regulation by α- and β-Adrenergic Receptors in Conscious Calves

S.F. Vatner[1], M.A. Young[1], and D.E. Vatner[2]

Summary. Regulation of coronary arteries by α- and β-adrenergic receptor subtype mechanisms was examined in chronically-instrumented, conscious calves. The physiological data on α- and β-adrenergic receptors were correlated with ligand binding studies from sarcolemmal membranes of calf large coronary arteries. In the conscious animal, selective α_1- + α_2- and β_1- + β_2-adrenergic agonists were examined in the presence and absence of selective α_1- + α_2-, and β_1- + β_2-adrenergic receptor blockades. Both the α_1-adrenergic agonist, phenylephrine, and the α_2-adrenergic agonist, B-HT 920, induced similar vasoconstriction of the large coronary arteries in the conscious calf, whereas the β_1-adrenergic agonist, prenalterol, and β_2-adrenergic agonist, pirbuterol, induced vasodilation. These effects were abolished by their respective adrenergic subtype blocker, but were not significantly affected by the corresponding blocker for the other subtype. Ligand binding studies demonstrated the presence of both subtypes of α- and β-adrenergic receptors, with a predominance of the β_1- and α_2-adrenergic receptor subtypes. Thus, large coronary arteries of the calf contain both α_1- + α_2- and β_1- + β_2-adrenergic receptor subtypes, and agonists are capable of eliciting significant constriction with either of the α-adrenergic subtypes, and dilation with either of the β-adrenergic subtypes.

Key Words: Ligand binding – Coronary artery diameter – Sympathetic nervous system – Acetylcholine – Adrenergic receptor

Introduction

The regulation of large coronary arteries by α and β-adrenergic receptor subtypes remains controversial [1–5]. Most earlier work has been conducted on isolated preparations due to the difficulties in studying large coronary arteries in the intact animal. The results of previous in vitro studies have suggested that large coronary arteries contain predominantly α_1-adrenergic [6–11] and β_1-

---

[1] Department of Medicine, Harvard Medical School, New England Regional Primate Research Center, Southborough, MA 01772, USA
[2] Department of Medicine, Children's Service, Harvard Medical School, Massachusetts General Hospital, Boston, MA 02114, USA

adrenergic [12–17] receptor subtypes, and consequently do not respond to α_2- or β_2-adrenergic stimulation. However, techniques have been developed, which permit the investigation of large coronary artery regulation in the intact, conscious animal [18–22]. These techniques, coupled with ligand binding analyses of adrenergic receptor subtypes, provide a novel approach to reconciling the controversy in this field. The goal of this article is to review the work conducted on the regulation of large coronary arteries in calves, which correlated physiological studies in intact, conscious calves, with ligand binding experiments examining receptor subtype composition in membrane preparations from bovine coronary arteries [18,22].

Materials and Methods

Physiological Studies [18,22]

Female calves, 6–10 weeks old, fully weaned, and weighing 60–80 kg, were anesthetized with halothane (0.5–1.5 vol% in oxygen) following pre-anesthesia with sodium thiamylal (10–15 mg/kg i.v.). Using sterile surgical technique, and with ventilation controlled by a Harvard respirator, a left thoracotomy was performed via the fourth intercostal space. Two miniature ultrasonic dimension transducers were implanted using 5–0 suture on opposing adventitial surfaces of the left circumflex coronary artery, 4–6 cm from its origin. A Doppler blood flow transducer and hydraulic constrictor were implanted in the same vessel. Care was taken during the implantation to avoid excessive dissection and possible damage to the vessel and perivascular nerves. Proximal to the dimension transducers, a silastic catheter was implanted in the left circumflex coronary artery. In addition, Tygon catheters were implanted in the descending aorta and left atrium. Animals used in this study were maintained in accordance with the guidelines of the Committee on Animals of the Harvard Medical School [23] and the National Institutes of Health's "Guide for Care and Use of Laboratory Animals" [24].

The experiments were conducted 1–3 weeks after surgery in healthy, conscious calves, lying quietly in a cart. Measurements of left circumflex coronary arterial diameter and aortic pressure were continuously recorded. Dose-response relationships to intracoronary selective α_1-adrenergic stimulation with B-HT 920 were examined in the presence of β-adrenergic receptor blockade with propranolol [25–29]. Dose-response relationships to intracoronary selective β_2-adrenergic stimulation with pirbuterol, and selective β_1-adrenergic stimulation with prenalterol were also examined [30–32,18,20–22]. On separate days, the effects of the α-adrenergic agonists were studied following selective α_1-adrenergic receptor blockade with prazosin, and α_2-adrenergic receptor blockade with rauwolscine, and the effects of β-adrenergic agonists were studied following selective β_1-adrenergic receptor blockade with atenolol or betaxolol, and following selective β_2-adrenergic blockade with ICI 118, 551 [18,22]. The data were analyzed using a paired t-test and Bonferroni correction [33].

Biochemical Studies [18,22]

After calves of both sexes were killed at the slaughter house, the hearts were quickly placed in iced saline and transported within 20 min to the laboratory. Sarcolemmal membranes were prepared from the large coronary arteries. All binding studies were performed in triplicate. ^{125}I-cyanopindolol (^{125}I-CYP) (0.025–1.0 nM) was used for β-adrenergic antagonist binding studies for Scatchard analysis. The percentage of $β_1$- and $β_2$-adrenergic receptor subtypes in the plasma membrane homogenate of calf large coronary artery smooth muscle was determined by competitive inhibition ligand binding studies with the nonselective antagonist radioligand ^{125}I-CYP (0.2 nM), and with multiple concentrations (10^{-9} M to 10^{-4} M) of the selective $β_1$-agonist prenalterol [30], with the selective $β_2$-agonist pirbuterol [32], with the selective $β_1$-antagonist betaxolol [34], and with the selective $β_2$-antagonist ICI-118, 551 [35]. The high-

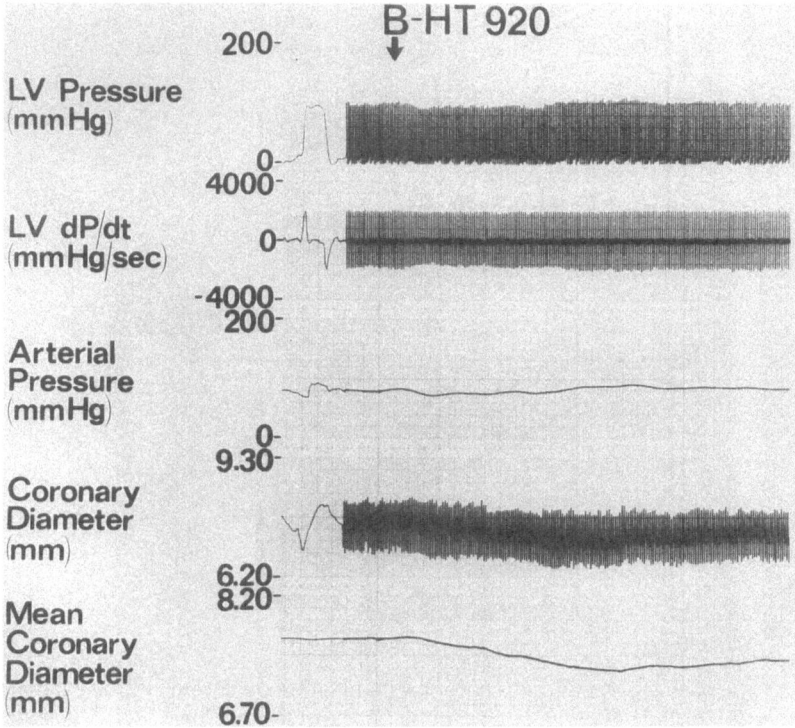

Fig. 1. Experimental record of left ventricular (*LV*) pressure, first derivative of LV pressure (*LV dP/dt*), mean arterial pressure, and coronary artery diameter in response to $α_2$-adrenoceptor activation with intracoronary bolus of B-HT 920 (5.0 µg/kg). B-HT 920 elicited significant constriction of epicardial coronary artery, with only minor changes in systemic hemodynamics. (From [22] with permission of the American Physiological Society)

and low-affinity binding constants (K_H and K_L) and the ratio of high- to low-affinity binding sites ($R_H:R_L$) were determined for each agonist and antagonist in the large coronary artery smooth muscle sarcolemma preparation by computer modeling, as described by Munson and Rodbard [36]. For α-adrenergic receptor studies determinations of binding of [³H]-prazosin, the $α_1$-adrenergic receptor specific antagonist, or [³H]-rauwolscine, the $α_2$-adrenergic receptor specific antagonist, to the membrane preparations were performed in triplicate. The concentration range of radioligand utilized in saturation binding studies was 0.5–30 nM for [³H]-rauwolscine, and 0.01–3 nM for [³H]-prazosin. These procedures have been previously described in detail [18,22].

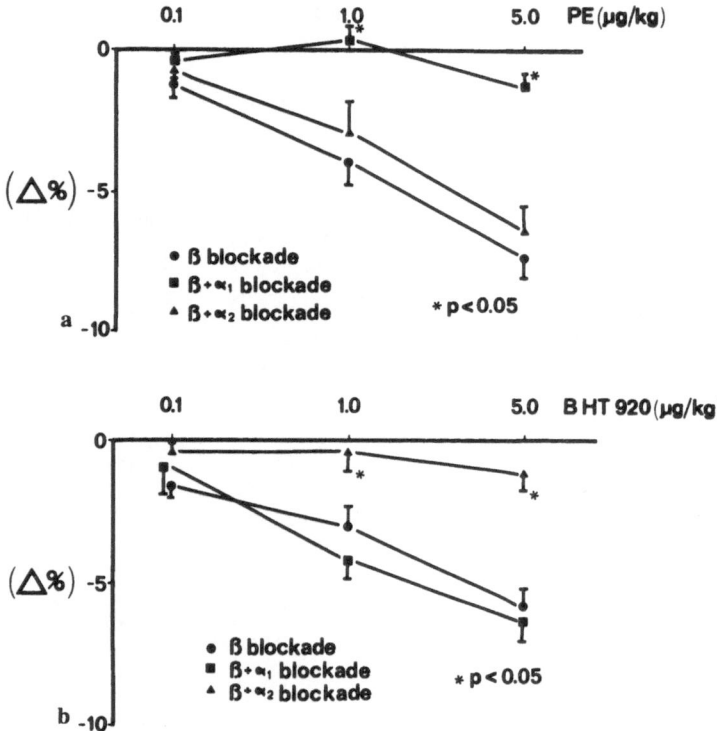

Fig. 2.a. Effects of $α_1$-selective adrenoceptor stimulation on coronary artery diameter in presence of β-blockade alone, and after either β + $α_1$-blockade or β + $α_2$-blockade. The fall in coronary artery diameter with phenylephrine (*PE*) was unchaged in the presence of β + $α_2$-blockade, but was abolished by $α_1$-blockade. Δ%, % change; *, Δ% in coronary diameter, which is significantly less than during β-blockade alone. **b** Effects of $α_2$-selective adrenoceptor stimulation on coronary artery diameter in presence of β-, β + $α_1$, or β + $α_2$ blockade. The fall in coronary artery diameter with B-HT 920 was unchanged after β + $α_1$-blockade, but it was abolished by $α_2$-blockade. (From [22] with permission of the American Physiological Society)

Results

Physiological Regulation by α-Adrenergic Subtype Mechanisms

In the conscious calf, intracoronary administration of both phenylephrine, a selective α_1-adrenergic agonist, and B-HT 920, a selective α_2-adrenergic agonist caused dose-related constriction, i.e., decreases in the diameter of the large coronary artery. An example of the constriction induced by B-HT 920 is shown in Fig. 1 [22]. Figure 2 [22] shows the average decreases in coronary diameter in response to phenylephrine and B-HT 920. The decreases in diameter were significantly attenuated by their subtype specific blockers, i.e., the large coronary vasoconstriction induced by phenylephrine was essentially blocked by prazosin and not affected by rauwolscine, whereas the constriction induced by B-HT 920 was essentially blocked by rauwolscine and not affected by prazosin. Norepinephrine, an α_1 and α_2 agonist also induced significant large coronary constriction, which was attenuated by selective α_1- or α_2-adrenergic receptor blockade, and abolished by combined α_1- and α_2-adrenergic receptor blockade (Fig. 3) [22].

Physiological Regulation by β-Adrenergic Subtype Mechanisms

In the conscious calf, both the intracoronary administration of the selective β_1-adrenergic agonist, prenalterol, and the selective β_2-adrenergic agonist, pirbuterol, and the combined β_1 and β_2-adrenergic agonist, isoproterenol, induced dose-related dilation of the large coronary arteries of the conscious calf. An example of the effects of isoproterenol in a conscious calf is shown in Fig. 4 [18] when blood flow was allowed to increase (Fig. 4a) and when blood flow was

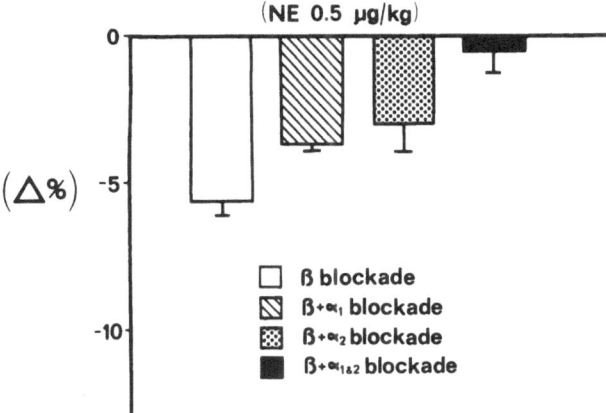

Fig. 3. Effects of norepinephrine (*NE*) on coronary artery diameter after β-, $\beta + \alpha_1$, $\beta + \alpha_2$-, or $\beta + \alpha_{1+2}$-blockade. The fall in coronary diameter with NE was attenuated by either α_1- or α_2-blockade, and completely abolished after combined α_{1+2}-blockade. $\Delta\%$, % change. (From [22] with permission of the American Physiological Society)

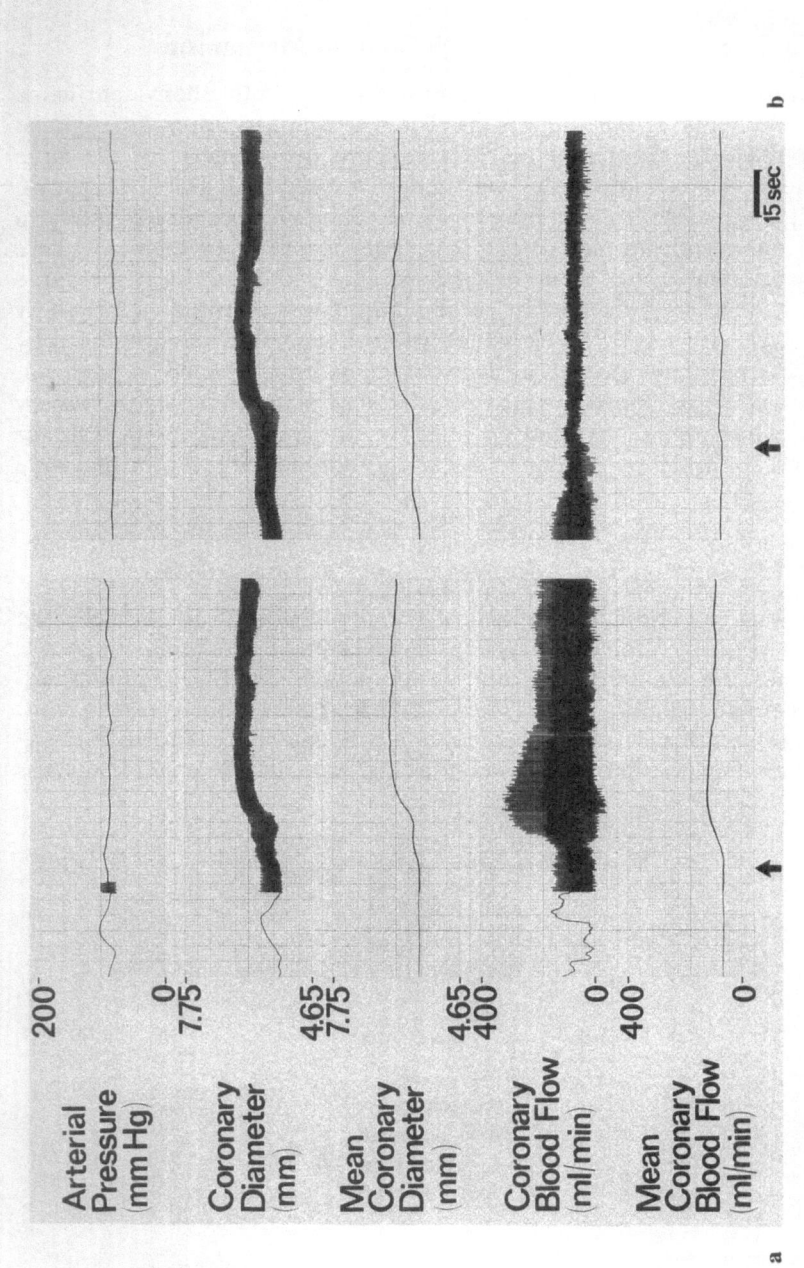

Fig. 4. The effects of intracoronary bolus injection of the mixed β_1- and β_2-adrenergic agonist isoproterenol (0.0025 µg/kg) on arterial pressure, phasic and mean coronary artery diameter, and phasic and mean coronary blood flow. **a** coronary blood flow was unrestricted; **b** when blood flow was restricted to pre-injection control levels with an hydraulic cuff constrictor, the rise in coronary artery diameter was unchanged, which indicates that dilation of the large artery is due to direct activation of β-receptors. (From [18] with permission of the American Heart Association)

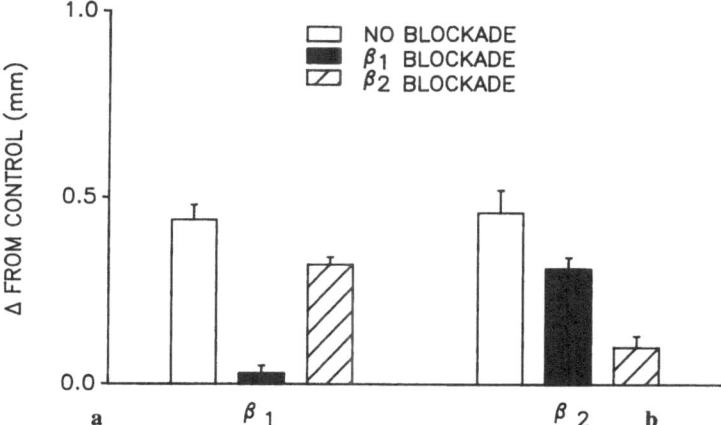

Fig. 5. The increases in the large coronary diameter are compared for **a** β₁-adrenergic stimulation with prenalterol and **b** β₂-adrenergic stimulation with pirbuterol, in the absence of blockade (*open bars*), after selective β₁-adrenergic blockade (*solid bars*), and after selective β₂-adrenergic blockade (*diagonal bars*). Both β₁- and β₂-adrenergic stimulation induced equivalent dilation of large coronary arteries in the conscious calf, which was blocked by their respective subtype blockers

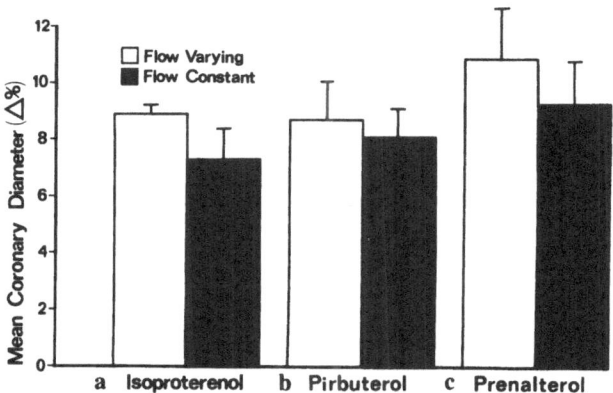

Fig. 6. The effects of **a** isoproterenol; **b** pirbuterol; **c** prenalterol are compared when coronary blood flow was allowed to increase (*open bars*), and when coronary blood flow was held constant (*solid bars*). The changes in coronary diameter with the drugs under both conditions were nearly identical, indicating that the vasodilation was independent of changes in blood flow. (From [18] with permission of the American Heart Association)

held constant (Fig. 4b). The effects of prenalterol (0.4 μg/kg) and pirbuterol (0.25 μg/kg) in the presence and absence of subtype specific blockers are shown in Fig. 5. The β₁-adrenergic dilation induced by prenalterol was essentially abolished by the β₁-adrenergic blocker atenolol, but not significantly affected by the β₂-adrenergic blocker, ICI 118, 551. In contrast, the β₂-adrenergic dilation induced by pirbuterol was essentially blocked by the β_2-adrenergic blocker, ICI 118, 551, but not significantly affected by the β_1-adrenergic blocker atenolol. Holding blood flow constant by constricting the coronary artery via the implanted hydraulic constricting device, and preventing the increase in blood flow due to distal resistance vessel dilation, did not significantly affect the vasodilation of the proximal large coronary arteries induced by intracoronary β_1- and β_2-adrenergic agonists (Fig. 6 [18]).

Ligand Binding Studies

The density of α- and β-adrenergic receptors was similar, ranging from 50–80 fmol/mg protein. Utilizing antagonist binding, the fraction of α_2-adrenergic receptors was approximately four times the number of α_1-adrenergic receptors (Fig. 7 [22]). The fraction of β_1- and β_2-adrenergic receptors was derived using both agonist and antagonist binding studies. Using isoproterenol, epinephrine, and norepinephrine competition curves, the calf coronary artery preparation demonstrated an intermediate pattern of β_1- and β_2-adrenergic subtypes (Fig. 8 [18]). A preponderance of β_1-adrenergic receptors was further substantiated by examining competition curves with the subtype-specific adrenergic agonists, prenalterol (β_1) and pirbuterol (β_2), and the subtype-specific antagonists, betaxolol (β_1) and ICI 118, 551 (β_2) (Fig. 9 [18]). Computer modeling consistently demonstrated that the β_1-adrenergic subtype was present in excess of the β_2-adrenergic subtype in calf coronary artery. The fraction of β_1-

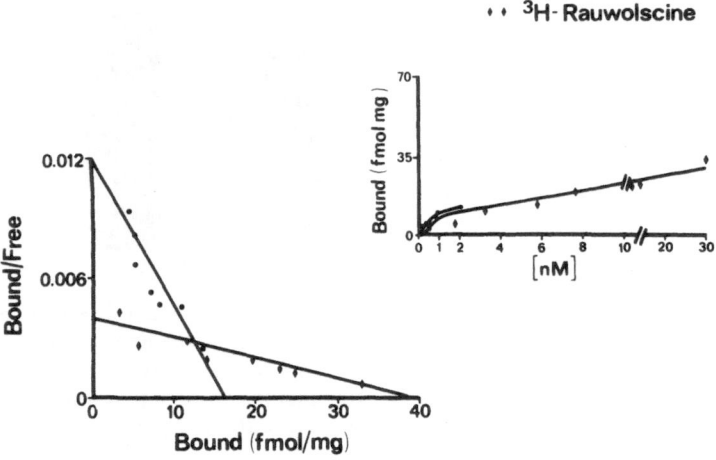

Fig. 7. Scatchard analysis of α_1-adrenoceptor binding of calf coronary artery smooth muscle membranes with [^{3}H] prazosin and α_2-adrenoceptor binding with [^{3}H] rauwolscine. Binding of [^{3}H] prazosin (0.65 ± 0.15 nM) has a higher affinity for α_1-adrenoceptors than that of [^{3}H] rauwolscine (7.35 ± 1.19 nM) for α_2-receptors, yet the number of α_1-receptors (15 ± 3.1 fmol/mg protein) is less than α_2-receptors (68 ± 5.1 fmol/mg protein). (From [22] with permission of the American Physiological Society)

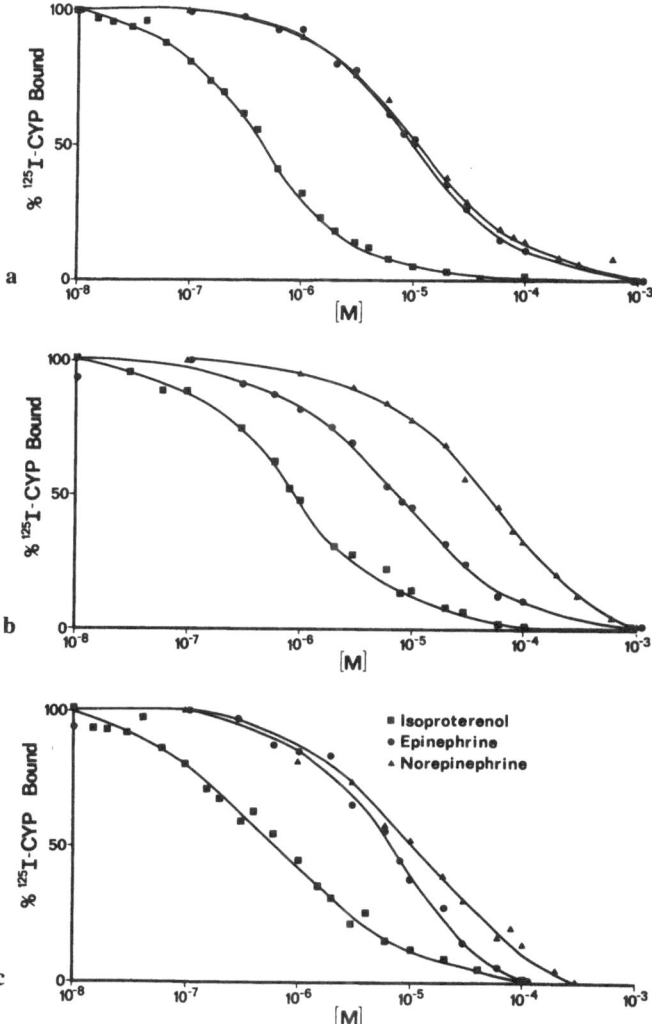

Fig. 8. Catecholamine competition curves are shown using calf heart purified sarcolemma with isoproterenol (*squares*), epinephrine (*circles*), and norepinephrine (*triangles*). **a** β_1 subtype rank order of potency is shown with isoproterenol > epinephrine = norepinephrine. **b** Competition curves are shown from calf lung membranes. This series of curves is consistent with the β_2 subtype rank order of potency, i.e., isoproterenol > epinephrine > norepinephrine. **c** Calf coronary artery sarcolemma was studied with catecholamine competition curves. Epinephrine and norepinephrine curves are closely aligned and suggest a β_1-subtype predominance in the coronary artery. (From [18] with permission of the American Heart Association)

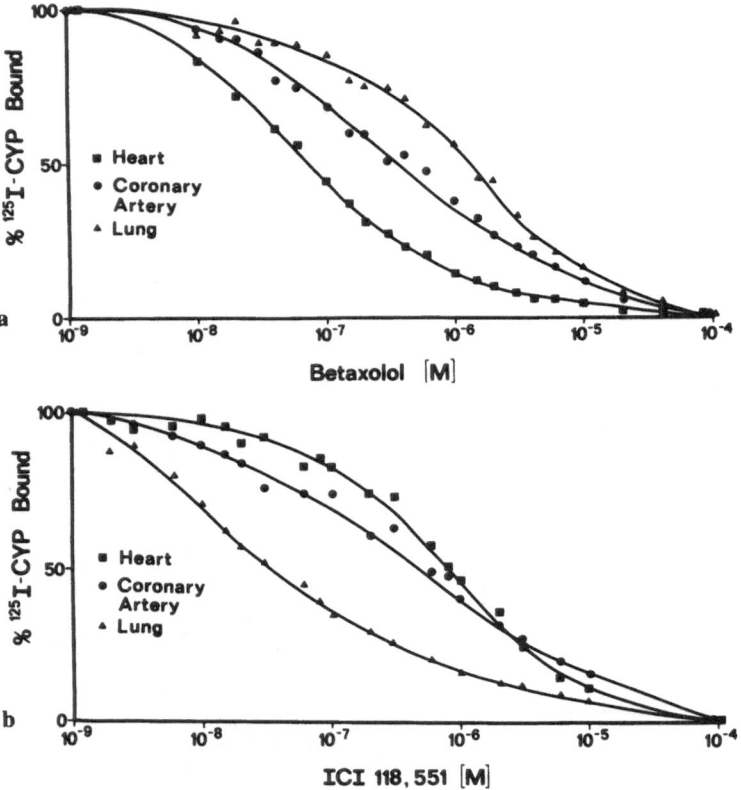

Fig. 9.a. The effects of increasing concentrations of betaxolol are shown on ^{125}I-CYP binding to heart (*squares*), coronary artery (*circles*), and lung membranes (*triangles*). The affinity of betaxolol was greatest for heart, least for lung, and intermediate for coronary artery. These data are consistent with the concept that the heart is mostly β_1, lung is mostly β_2, and coronary artery demonstrated a mixture of β subtypes. (From [18] with permission of the American Heart Association). **b** ICI 118, 551, a β_2-selective antagonist was used in increasing concentrations with a constant concentration of ^{125}I-CYP (0.2 nM). The affinity of ICI 118, 551 was greatest for lung (*triangles*), least for heart (*squares*), and intermediate for coronary artery (*circles*). These data are consistent with the concept that coronary artery has a mixture of β_1- and β_2-adrenergic receptor subtypes. (From [18] with permission of the American Heart Association)

adrenergic receptors was approximately 75% of the total receptors; β_2-adrenergic receptors comprised approximately 25% of the total receptors.

Discussion

A unique feature of the experiments reviewed in this paper is the combination of functional and biochemical approaches to the problem of adrenergic receptor subtype distribution in large coronary arteries [18,22]. Another unique feature is that the functional studies were conducted in intact, conscious animals, in contrast to most of the previous work, which had been conducted in vitro or on anesthetized animals. These earlier studies examined large coronary arteries, and concluded that α-adrenergic constriction was mediated exclusively by the $α_1$-adrenergic subtype [6–11], whereas β-adrenergic dilation of large coronary arteries was thought to be mediated predominantly by the $β_1$-adrenergic subtype [12–17]. The results, obtained by utilizing the combined approach of both functional and biochemical analyses [18,22], indicate that earlier in vitro demonstrations of only $α_1$- and $β_1$-adrenergic receptors are not applicable to the intact, conscious calf. Instead, it appears that $α_1$- and $α_2$-, and $β_1$- and $β_2$-adrenergic receptors are all present in bovine large coronary arteries, and that both α-adrenergic receptor subtypes are capable of eliciting vasoconstriction, and both β-adrenergic receptor subtypes are capable of eliciting vasodilation. The conclusion that both α-adrenergic subtypes are capable of eliciting large coronary vasoconstriction is based on the data that the selective $α_1$-agonist (phenylephrine) and selective $α_2$-agonist (B-HT 920) produce equivalent, dose-dependent reductions in large coronary diameter, and that the large coronary artery constrictions with phenylephrine and B-HT 920 are abolished by the selective $α_1$- and $α_2$-adrenergic antagonists, prazosin or rauwolscine, respectively [22]. These conclusions are further strengthened by ligand binding studies, which demonstrate the presence of both $α_1$- and $α_2$-adrenoreceptors in coronary sarcolemmal membranes. Surprisingly, in view of the prior work in this field, the predominant subtype identified by ligand binding was the $α_2$-adrenergic receptor. However, in our studies it could not be discerned whether the $α_2$-adrenergic receptors were pre- or postjunctional.

The physiological studies presented also indicate that the large coronary arteries of the calf contain both $β_1$- and $β_2$-adrenergic receptors [18]. This conclusion is based on finding similar increases in large coronary artery diameter in response to prenalterol, which stimulates primarily $β_1$-adrenergic receptors, and pirbuterol, which stimulates primarily $β_2$-adrenergic receptors. Furthermore, the preferential $β_1$- (atenolol, betaxolol) and $β_2$- (ICI 118,551) adrenergic receptor antagonists essentially abolished the responses to their respective agonists, but did not abolish responses to the other agonist. While prior studies in conscious dogs have suggested that both $β_1$- and $β_2$-adrenergic receptor stimulation and blockade can affect large coronary arteries [20,21], those studies were not conclusive since there were concomitant changes in coronary blood flow, and the agonists and antagonists were administered systemically rather than directly into the coronary artery. In the studies in conscious calves [18], the agonists were administered directly in the coronary arteries. Furthermore, equivalent $β_1$- and $β_2$-adrenergic dilation was observed whether or not blood flow was held constant (Fig. 3). This was not predicted, since increases in blood flow, per se, can induce dilation of large coronary arteries [37]. However, this mechanism was not of great importance in mediating the large coronary arterial dilation in response to $β_1$- and $β_2$-adrenergic activation.

In summary, large coronary arteries of the calf contain both α_1- + α_2- and β_1- + β_2- adrenergic receptor subtypes. Selective adrenergic agonists are capable of eliciting significant constriction with either of the α-subtypes and dilation with either of the β-subtypes in the conscious calf.

Acknowledgements. This work was supported in part by U.S. Public Health Service Grants HL 38 070, HL 33 107, and RR 00 168.

References

1. Berne RM, Rubio R (1979) Coronary circulation. In: Geiger SR (ed) Handbook of physiology, sect 2: The cardiovascular system, vol 1. American Physiological Society, Washington, DC, pp 873–952
2. Feigl EO (1983) Coronary physiology. Physiol Rev 63: 1–205
3. Marcus MD (1983) The coronary circulation in health and disease. McGraw Hill, New York
4. Young MA, Vatner SF (1986) Brief review: Regulation of large coronary arteries. Circ Res 59: 579–595
5. Young MA, Knight DR, Vatner SF (1987) Autonomic control of large coronary arteries and resistance vessels. Prog Cardiovasc Dis 30: 211–234
6. Cocks TM, Angus JA (1983) Endothelium-dependent relaxation of coronary arteries by noradrenaline and serotonin. Nature 305: 627–630
7. Cohen RA, Shepherd JT, Vanhoutte PM (1983) Prejunctional and postjunctional actions of endogenous norepinephrine at the sympathetic neuroeffector junction in canine coronary arteries. Circ Res 52: 16–25
8. Heusch G, Deussen A, Schipke J, Thaemer V (1984) α_1- and α_2-adrenoceptor-mediated vasoconstriction of large and small canine coronary arteries in vivo. J Cardiovasc Pharmacol 6: 961–968
9. Kaumann AJ (1983) Yohimbine and rauwolscine inhibit 5-hydroxytryptamine-induced contraction of large coronary arteries of calf through blockade of 5 HT_2 receptors. Naunyn Schmiedebergs Arch Pharmacol 323: 149–154
10. Rimele TJ, Rooke TW, Aarhus LL, Vanhoutte PM (1983) α_1-adrenoceptors and calcium in isolated canine coronary arteries. J Pharmacol Exp Ther 226: 668–672
11. Toda NT, Okamura T, Nakajima M, Miyazaki M (1984) Modification by yohimbine and prazosin of the mechanical response isolated dog mesenteric, renal and coronary arteries to transmural stimulation and norepinephrine. Eur J Pharmacol 98: 69–75
12. Baron GD, Speden RN, Bohr DF (1972) Beta-adrenergic receptors in coronary and skeletal muscle arteries. Am J Physiol 223: 878–881
13. Cornish EJ, Miller RC (1975) Comparison of the β-adrenoceptors in the myocardium and coronary vasculature of the kitten heart. J Pharm Pharmacol 27: 23–30
14. de la Lande IS, Harvey JA, Holt S (1974) Response of the rabbit coronary arteries to autonomic agents. Blood Vessels 11: 319–337
15. Drew GM, Levy GP (1972) Characterization of the coronary vascular β-adrenoceptor in the pig. Br J Pharmacol 46: 348–350
16. Johansson B (1973) The β-adrenoceptors in the smooth muscle of pig coronary arteries. Eur J Pharmacol 24: 218–224
17. Schwartz J, Velly J (1983) The β-adrenoceptor of pig coronary arteries: Determination of β_1 and β_2 subtypes by radioligand binding. Br J Pharmacol 79: 409–414
18. Vatner DE, Knight DR, Homcy CJ, Vatner SF, Young MA (1986) Subtypes of beta adrenergic receptors in bovine coronary arteries. Circ Res 59: 463–473

19. Vatner SF, Pagani M, Manders WT, Pasipoularides AD (1980) Alpha adrenergic vasoconstriction and nitroglycerin vasodilation of large coronary arteries in the conscious dog. J Clin Invest 65: 5–14
20. Vatner SF, Hintze TH, Macho P (1982) Regulation of large coronary arteries by β-adrenergic mechanisms in the conscious dog. Circ Res 51: 56–66
21. Vatner SF, Hintze TH (1983) Mechanisms of constriction of large coronary arteries by beta-adrenergic receptor blockade. Circ Res 53: 389–400
22. Young MA, Knight DR, Homcy CJ, Graham RM, Vatner SF, Vatner DE (1988) α-adrenergic vasoconstriction and receptor subtype in large coronary arteries of calves. Am J Physiol 255 (Heart Circ Physiol 24): H1452–H1459
23. Harvard Medical School (1978) Guidelines for animal care and use of laboratory animals (Internal Document), Boston, MA
24. Guide for care and use of laboratory animals (1985) DHHS Publication NO Bethesoa, MD [NIH] 83–23
25. Kobinger W, Pichler L (1981) α_1- and α_2-adrenoceptor subtypes: selectivity of various agonists and relative distribution of receptors as determined in rats. Eur J Pharmacol 73: 313–321
26. Van Zweiten PA, Timmermans PBMWM (1983) Cardiovascular α_2-adrenoceptors. J Mol Cell Cardiol 15: 717–733
27. Woodman OL, Vatner SF (1987) Coronary vasoconstriction mediated by α_1- and α_2-adrenoceptors in the conscious dog. Am J Physiol 253: H388–H393
28. Woodman OL, Constantine JW, Vatner SF (1986) Nifedipine attenuates both α_1- and α_2-adrenoceptor mediated pressor and vasoconstrictor responses in conscious dogs and primates. J Pharmacol Exp Ther 239: 648–653
29. Woodman OL, Vatner SF (1986) Cardiovascular responses to the stimulation of α_1- and α_2-adrenoceptors in the conscious dog. J Pharmacol Exp Ther 237: 86–91
30. Carlsson E, Dahlof CG, Hedberg A, Persson H, Tangstrand B (1977) Differentiation of cardiac chronotropic and inotropic effects of β-adrenoceptor agonists. Naunyn Schmiedebergs Arch Pharmacol 300: 101–105
31. Manders WT, Vatner SF, Braunwald E (1980) Cardio-selective beta-adrenergic stimulation with prenalterol in the conscious dog. J Pharmacol Exp Ther 215: 266–270
32. Moore PF, Constantine JW, Barth WE (1978) Pirbuterol, a selective β_2-adrenergic bronchodilator. J Pharmacol Exp Ther 207: 410–418
33. Miller RG (1966) Simultaneous statistical inference. McGraw-Hill, New York, pp 116–126
34. Cavero I, Lefevre-Borg F, Manoury P, Roach AG (1983) In vitro and in vivo pharmacological evaluation of betaxolol, a new, potent, and selective β_1-adrenoceptor antagonist. In: Morselli PL (ed) Laboratories d'etudes et de recherches synthelabo. Raven, New York, pp 31–42
35. Bilski A, Dorries S, Fitzgerald JD, Jessup R, Tucker H, Wale J (1979) ICI118,551: A potent $beta_2$-adrenoceptor antagonist. Br J Pharmacol 69: 292P–293P
36. Munson PJ, Rodbard D (1980) LIGAND: A versatile computerized approach for characterization of ligand-binding systems. Anal Biochem 107: 220–239
37. Hintze TH, Vatner SF (1984) Reactive dilation of large coronary arteries in conscious dogs. Circ Res 68: 50–57

CHAPTER 4

Effects of Native and Oxidized Low Density Lipoproteins on Formation and Inactivation of EDRF and Vascular Smooth Muscle

E. Bassenge and J. Galle[1]

Summary. Native and oxidized low density lipoproteins (LDL) were investigated for their direct influence on endothelium-derived relaxing factor (EDRF)-formation, EDRF-activity, and vascular smooth muscle tone. Native (n) LDL was isolated from fresh human plasma via sequential ultracentrifugation, and oxidized by Cu^{2+}-incubation. EDRF released from cultured endothelial cells was inactivated by both n-LDL and ox-LDL (1 mg/ml), as detected in a bioassay system. n-LDL reduced the EDRF-mediated vasodilations of the detector segments by $38.5 \pm 5.3\%$, and ox-LDL by $55.5 \pm 4.6\%$. The effects of lipoproteins on EDRF-formation were studied on cultured endothelial cells, preincubated with either n-LDL or ox-LDL (1mg/ml for 1 hour) and stimulated for EDRF-release with bradykinin *after* washout of the lipoproteins. EDRF was assessed by measuring its stimulatory effect on the activity of a purified soluble guanylate cyclase. Preincubation with both n-LDL and ox-LDL did not reduce the bradykinin-induced EDRF-formation. Accordingly, acetylcholine-induced, EDRF-mediated dilations of intact rabbit femoral artery segments were not impaired by luminal exposure to n-LDL or ox-LDL (1mg/ml for 1 hour).

Effects of n-LDL and ox-LDL on vascular smooth muscle tone were investigated in isolated perfused rabbit femoral arteries. Perfusion of endothelium-intact and endothelium-denuded segments with ox-LDL (80–500 µg protein/ml) caused no, or only weak vasoconstriction in the absence of contractile agonists. However, in the presence of ox-LDL, vasoconstrictions to threshold concentrations of norepinephrine (NE), serotonin (5-HT), phenylephrine (PE) and potassium were significantly enhanced. Native LDL (80–1000 µg/ml) had no effect on vascular tone, neither in the presence nor absence of contractile agonists. Incubation with verapamil, diltiazem and nitrendipine inhibited vasoconstrictions evoked by ox-LDL. The contractile responses to ox-LDL were significantly greater in endothelium-denuded segments than in endothelium-intact segments.

In conclusion, neither n-LDL nor ox-LDL acutely impair the formation of EDRF, but inactivate EDRF after its release from endothelial cells. n-LDL has no direct influence on vascular smooth muscle tone, but ox-LDL greatly enhances vasoconstrictions to various contractile agonists by direct interaction with vascular smooth muscle. Thus, in

---

[1] Department of Applied Physiology, University of Freiburg, 7,800 Freiburg, Federal Republic of Germany

regions of lipoprotein accumulation in the arterial wall, the inactivation of EDRF and the potentiation of agonist-induced vasoconstrictions may favor inappropriate vasoconstrictions.

Introduction

The atherogenic properties of low density lipoproteins (LDL) are well established, and recent evidence suggests that oxidation of LDL is an important step in atherogenesis [1]. However, it remains controversial, whether lipoproteins affect directly arterial endothelial function or smooth muscle tone. Several studies have shown impairment of endothelium-mediated vasodilation in hypercholesterolemia in animals and humans [2–5], and arterial hyperresponsiveness to different vasoconstrictors in hypercholesterolemic animals [2, 6–10]. In in vitro investigations, it was found that both n-LDL and ox-LDL can be cytotoxic to endothelial cells [11–13], and recently it was proposed that high concentrations of LDL inhibit EDRF-formation via the endothelial LDL-receptor [14].

In hypercholesterolemia, LDL is found in the arterial wall between the endothelium and the vasculature [15], and strong evidence also suggests that oxidized LDL accumulates in this region [16–18]. Since vascular tone represents a net balance between vasoconstrictor and vasodilator influences [19], we investigated the effects of LDL and its oxidized derivative on transfer and formation of EDRF, and on vascular smooth muscle tone.

Material and Methods

Preparation of Native LDL

Plasma was separated from freshly drawn human blood, and ethylene-diamine-tetraacetic acid (EDTA, 0.2 mM), butylated-hydroxy-toluene (BHT, 20 µM), chloramphenicol (10 mg/dl), and phenyl-methyl-sulphonyl-fluoride (PMSF, 1 mM) were added to avoid autoxidation, proteolytic digestion, and bacterial growth [20]. LDL (1.019–1.063 g/ml) was isolated by sequential ultracentrifugation at 200,000 g [21], and concentrated and purified by gel filtration in the presence of EDTA 0.2 mM and BHT 20 µM. Stock solutions of LDL (8–12 mg protein/ml, dissolved in Tyrode's solution) were sterilized by filtration (millex, Millipore, pore size 0.22 µm), and kept in the dark at 4°C for no longer than 3 weeks. LDL prepared by this method is refered to as native-LDL (n-LDL). Protein content was measured as described by Bradford [22].

Oxidative Modification of LDL

n-LDL was separated from EDTA and BHT by gel filtration. LDL (0.3 mg protein/ml) free of added antioxidants was incubated with $CuSo_4$ (5 µM) for 24 h at 23°C in phosphate-buffered saline. The degree of oxidation in comparison to n-LDL was quantified by three different methods: 1) the absorption at $\lambda = 234$ nm increased, indicating an increase in diene formation of fatty acids [23], 2) the relative mobility on agarose gel increased to up to 1.60 times of that of n-LDL, indicating an enhanced negative charge of ox-LDL [24], and 3) sodium-dodecyl-sulfate polyacrylamide gel electrophoresis (SDS-PAGE) demonstrated fractionation of apoprotein B 100 [24]. Finally, ox-LDL was separated from $CuSO_4$ by gel filtration, and concentrated as n-LDL.

Electrophoresis

Sodium-dodecyl-sulfate polyacrylamid gel electrophoresis (SDS-PAGE) was performed using 3% – 20% gradient gels, and xanthine oxidase, lipoxidase, albumine and peroxidase as molecular weight markers (283, 96, 68, and 40 kd, respectively) [25]. Visualization of the protein bands was achieved by coomassie blue [25]. n-LDL migrated as a single, apoB 100-protein band. ox-LDL was completely depleted of apoB 100-protein, and fractionated into multiple protein bands of less than 283 kd. Agarose gel electrophoresis was performed to measure relative mobility of ox-LDL compared to n-LDL, as index for lipoprotein oxidation [24]. Electrophoresis kits (lipidophor) were purchased from IMMUNO GmbH (Heidelberg, FRG).

Drugs

Phenylephrine, indomethacin, serotonin, bradykinin, thimerosal, and verapamil were purchased from Sigma (Munich, FRG). Indomethacin was dissolved in ethanol-0.1 M $NaHCO_3$ (1:3) vol/vol, and diluted with Tyrode's solution. Norepinephrine (Arterenol) (Hoechst, Frankfurt, FRG), diltiazem hydrochloride (Dilzem) (Goedecke, Berlin, FRG), and acetylcholine hydrochloride (Sigma, Munich, FRG) were also dissolved in Tyrode's solution (for composition see vessel preparation). For stock solutions of potassium-rich Tyrode's solution, sodium was isotonically replaced with potassium to gain a potassium concentration of 140 mM. Nitrendipine (Bayer, Leverkusen, FRG) was dissolved in absolute ethanol, and further diluted with Tyrode's solution. To avoid inactivation of nitrendipine by light, the drug and the stock solution were kept in the dark, and the perfusion routes of the experimental set up were covered by aluminium foil.

Bioassay Experiments

Endothelial cell culture. Bovine aortic endothelial cells were isolated and cultured as described elsewhere in detail [26]. The endothelium was scraped off freshly obtained aortae, and grown on standard culture dishes. For bioassay experiments with endothelium-denuded artery segments, cells were subcultured on microcarrier beads (Biosilon, Nunc-Intermed, Wiesbaden, FRG), and packed into a column. The endothelial cell column was perfused with oxygenated Tyrode's solution (PO_2 approximately 140 mmHg) at a rate of 30 ml/h, and

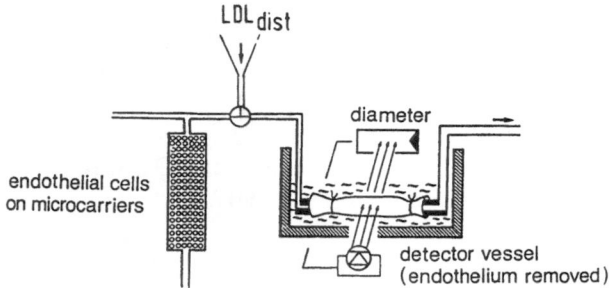

Fig. 1. Experimental setup used for bioassay of EDRF released from cultured endothelial cells after stimulation with thimerosal (5 μM). A preconstricted, endothelium-denuded detector vessel was perfused either by the EDRF containing effluent of an endothelial cell column, or by Tyrode's solution. Lipoproteins could be administered distally to the endothelial cell column. External diameter of the detector segment was recorded by a photoelectrical device.

stimulated for continuous EDRF-release with thimerosal (5 μM)[27]. The outflow tubing of the cell column was connected with the inflow cannula of a constricted artery segment which served as detector for EDRF (Fig. 1). As in earlier experiments [26], the dilator compound released from the cell columns was identified as EDRF. To investigate whether lipoproteins inactivate *released* EDRF, the lipoproteins were distally applied to the endothelial cell columns by a T-connection (Fig. 1). The perfusion rate through this sideway was $\frac{1}{10}$ of the perfusion rate through the endothelial cell columns, and the lipoproteins (stock solution concentration: 10 mg/ml) were administered through this sideway to give a final concentration of 1 mg/ml.

While testing the inhibitory action of lipoproteins, a second detector segment not perfused with the lipoproteins served as a control for continuous and stable EDRF-release. Therefore, the effluent of the endothelial cell columns could be perfused alternatively through the segments.

Inactivation of EDRF by lipoproteins is expressed as percent of the vasodilation in the absence of lipoproteins.

Guanylate cyclase assay. EDRF-release from cultured endothelial cells was detected by a guanylate cyclase assay as described elsewhere in detail [28, 29]. Endothelial cells grown on plates were exposed to n-LDL, ox-LDL, or lipoprotein-deficient serum as a control (1 mg/ml, 1h, respectively). The cells were then washed and stimulated with bradykinin (30 nM). The supernatant covering the cells was transfered to a test tube containing purified soluble guanylate cyclase from bovine lung. The activity of guanylate cyclase was determined by measuring the formation of $[^{32}p]$ cGMP from $[\alpha\text{-}^{32}P]GTP$.

Vessel preparation. Intact segments of the femoral artery were obtained from rabbits of either sex (2.5–3.5 kg). For part of the experiments, the endothelium was removed mechanically by gently rubbing the segments over a rough steel cannula. The intactness or absence of endothelium was tested as described earlier [30]. The segments were fixed between two steel cannulae, and placed in an organ bath (37°C) containing oxygenated Tyrode's solution (pH 7.4). The solution was perfused through the organ bath at a rate of 0.66 ml/min. In addition to organ bath perfusion, the segments were perfused intraluminally (Tyrode's solution: PO_2 130 mm Hg, PCO_2 28 mmHg, pH 7.38, at a rate of 0.5 ml/min).

Outer vascular diameters were continuously recorded by a photoelectric device. The transmural pressure was adjusted hydrostatically to 40 mmHg (isobaric conditions). Resting diameter under these conditions was 1755 ± 22 μm ($n = 28$). Full details of this experimental set-up (Fig. 1.) have been published earlier [30].

Endothelium-dependent vasodilations before and after 1 h incubation with n-LDL or ox-LDL (1 mg/ml) were elicited by intraluminal perfusion with cumulative dosages of acetylcholine ($0.03 - 1$ μM).

For bioassay experiments, the segments were preconstricted with norepinephrine (0.1 μM) applied to the organ bath, and the effluent of the endothelial cell columns was perfused intraluminally through the endothelium-denuded segments, as described above. Since ox-LDL has an enhancing effect on norepinephrine-induced vasoconstrictions [31], the norepinephrine concentration in the organ bath was lowered in presence of ox-LDL to achieve the same level of preconstriction as under reference conditions without ox-LDL.

Effects of the lipoproteins on smooth muscle tone were tested in each experiment in one pair of vascular segments, with and without endothelium, simultaneously mounted in two separate organ bath chambers. After an incubation period of 60 min, perfusion was started with Tyrode's solution. n-LDL (50–500 μg), ox-LDL (50–500 μg/ml), acetylcholine (Ach, 0.5 μM), indomethacin (10 μm), verapamil (1 μM), diltiazem (10 μM) and nitrendipine (1 μM) were administered at the intimal side by adding to the intraluminal perfusate. Norepinephrine (NE, 1–100 nM), phenylephrine (PE, 0.01–1 μM), serotonin (5-HT, 0.01–1 μM) and potassium (K^+, 20–40 mM) were added to the organ bath (if not otherwise indicated). For experiments with threshold concentrations of the contractile agonists, concentrations were chosen which evoked no or minimal vasoconstriction. Concentration-response relationships of NE and PE, with and without ox-LDL, were determined by adding the compounds to the intraluminal perfusate.

In an additional series of experiments, we compared the influence of n-LDL on endothelium-dependent vasorelaxations in arterial ring preparations with those in arterial segments. Pairs of rings and segments were prepared from the same rabbit aorta. The aortic segments were investigated in a similar manner to the femoral segments. The aortic rings (4 mm in diameter, 3 mm wide) were mounted in a force transducer (Biegestab K 30, H. Sachs, Hugstetten, FRG) under 1.5 g resting tension, and superfused with oxygenated Tyrode's solution (95% O_2–5% CO_2, pH 7.4, 37°C, 40 ml/h). After 30 min equilibration time, the rings were precontracted with norepinephrine (0.1 μM), and endothelium-dependent relaxations were elicited by 1 μM acetylcholine before and after 30 min superfusion with n-LDL (1 mg/ml) or Tyrode's solution without n-LDL (control). Relaxations are expressed as percent of precontraction. In each experiment, aortic rings and segments obtained from the same animal were investigated simultaneously.

Statistics

All data are presented as means ± SEM. Paired Student's *t*-test was used to evaluate statistical significance of differences. For multiple comparison of data, Bonferroni's correction was performed.

$P \leq 0.05$ was considered to be statistically significant. * = $P < 0.05$

Results

Formation of EDRF in cultured endothelial cells after incubation with n-LDL and ox-LDL

Figure 2 shows the release of EDRF from cultured endothelial cells after stimulation with bradykinin (30 nM). Prior to stimulation, endothelial cells were incubated with either n-LDL, ox-LDL, or lipoprotein-deficient serum as control (1 mg/ml for 1 hour respectively). Following bradykinin-stimulation of the endothelial cells, the guanylate cyclase (GC) activity in the assay increased more than threefold in all groups. The increase in GC-activity did not differ between the groups.

Inactivation by n-LDL and ox-LDL of EDRF Released from Cultured Endothelial Cells

Figure 3a illustrates the EDRF-inactivating effect of n-LDL (1 mg/ml) on a detector segment preconstricted with norepinephrine. The EDRF-mediated vasodilation of the detector segment was markedly reduced when n-LDL was added to the effluent of an endothelial cell column. The same holds true for administration of ox-LDL. Figure 3b illustrates the time course of the EDRF-inactivation by n-LDL during continous EDRF-perfusion of a detector

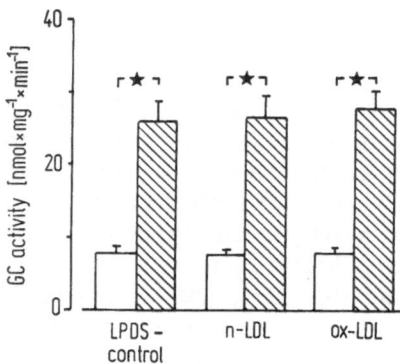

Fig. 2. Effects of incubation with n-LDL and ox-LDL (1 mg/ml for 1 hour) on bradykinin (*BK*)-induced EDRF-release from cultured endothelial cells. EDRF-release was measured as stimulation of the activity of guanylate cyclase (*GC*). Stimulation of the cells with bradykinin (30 nM, *hatched columns*) caused a more than threefold increase of basal GC-activity (*open columns*) in the control group (*left pair of columns*) as well as in the lipoprotein-incubated groups (*middle* and *right pair of columns*). LPDS, lipoprotein-deficient serum (control); *n* = 8; * $P < 0.05$

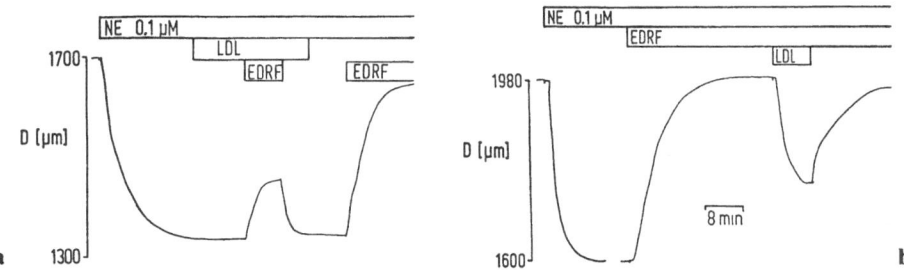

Fig. 3a,b. Diameter (D) recordings of an endothelium-denuded rabbit femoral artery segment, perfused with either Tyrode's solution, or the EDRF-containing effluent from an endothelial cell column stimulated with thimerosal (5 µM). The detector segment was preconstricted with norepinephrine 0.1 µM, and vasodilations were elicited by switching the intraluminal perfusate from Tyrode's solution to the EDRF containing effluent from the endothelial cell column. **a** In the presence of n-LDL (1 mg/ml) added to the effluent *distally* to the endothelial cell column, EDRF-mediated vasodilation was markedly reduced. **b** Adding n-LDL to the EDRF-containing effluent caused *immediate* suppression of vasodilation, which was also immediately reversible

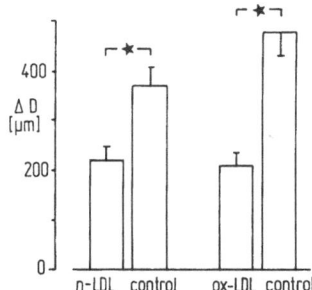

Fig. 4. Effects of 1 mg/ml n-LDL and ox-LDL on EDRF released from endothelial cell columns on its transfer to endothelium-denuded detector vessels. Addition of n-LDL to the effluent from endothelial cell columns reduced the vasodilations of the detector segments by 38.5 ± 5.3% (*left pair of columns*), and addition of ox-LDL reduced the vasolidations by 55.5 ± 4.6% (*right pair of columns*). $n = 12$;* $P < 0.05$

segment. Addition of n-LDL to the effluent caused immediate suppression of vasodilation, which was also immediately reversible when the n-LDL perfusion was stopped.

Data of the EDRF-inactivation by n-LDL and ox-LDL are summarized in Fig. 4: addition of n-LDL reduced the vasodilations of the detector segments by 38.5 ± 5.3%, and addition of ox-LDL by 55.5 ± 4.6% (n = 12).

Endothelium-Mediated Dilatation in Perfused Femoral Segments Preincubated with n-LDL and ox-LDL

Dose-response relations of preconstricted rabbit femoral arteries to acetylcholine before and after 1 h incubation with n-LDL or ox-LDL (1 mg/ml) are shown in Figs. 5a and 5b. Throughout the whole concentration range, dilation to acetylcholine did not differ significantly between lipoprotein-incubated segments and control segments. Thus, n-LDL and ox-LDL did not impair acetylcholine induced endothelium-dependent vasodilation in perfused segments.

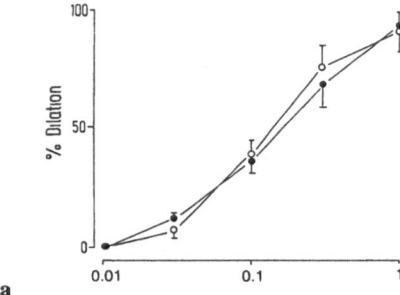 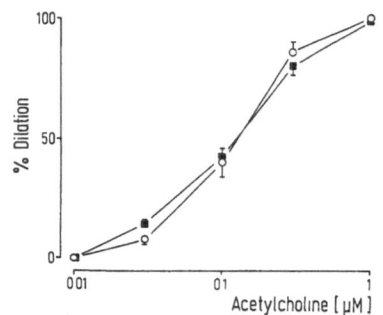

a   b
Acetylcholine [µM]

Fig. 5a,b. Vasodilations evoked by acetylcholine in intact rabbit femoral artery segments prior to and after 1 h incubation with 1 mg/ml n-LDL or ox-LDL. Acetylcholine was perfused intraluminally through the segments in cumulative dosages (0.03, 0.1, 0.3, 1 µM). Vasodilations are expressed as percent of preconstriction (induced by norepinephrine 0.1 µM). **a** Vasodilations before (control, -o-), and after incubation with n-LDL (-●-); $n = 7$. **b** Vasodilations before (control, -o-), and after incubation with ox-LDL (-■-); $n = 9$. The vasodilations did not differ significantly

Comparison of the Influence of n-LDL on Endothelium-Dependent Relaxations in Aortic Rings and Aortic Segments

As in femoral segments, vasodilation of aortic segments to acetylcholine (1 µM) was not impaired after 1 h incubation with 1 mg/ml n-LDL: 81 ± 5% dilation before vs 79 ± 5% dilation after incubation with n-LDL ($n = 4$). In contrast, relaxation of aortic rings to acetylcholine (1 µM) was markedly reduced after 30 min superfusion with n-LDL: 81 ± 8% relaxation vs 36 ± 15% relaxation ($n = 4$). Relaxation of control rings without incubation with lipoproteins did not change within this time interval.

Effects of n-LDL and ox-LDL on Unstimulated Segments.

Intraluminal perfusion of unstimulated segments (no contractile agonists present in the organ bath) with n-LDL in concentrations of 50–500 µg protein/ml did not elicit any vasoconstriction. Ox-LDL caused no, or only weak vasoconstriction on (17 ± 4 µm, max. 2% of resting diameter, $n = 26$) at concentrations below 150 µg protein/ml. When ox-LDL was administered at higher concentrations, moderate vasoconstriction (139 ± 11 µm, max. 8% of resting diameter, $n = 16$) were observed.

Effects of n-LDL and ox-LDL on Stimulated Segments

As shown by the diameter recording in Fig. 6, ox-LDL caused markedly augmented contractile responses in the presence of the contractile agonist phenylephrine, which was administered at threshold concentrations.

Figure 7 demonstrates that this potentiating effect of ox-LDL also takes place with other contractile agonists: contractile responses to norepinephrine (NE),

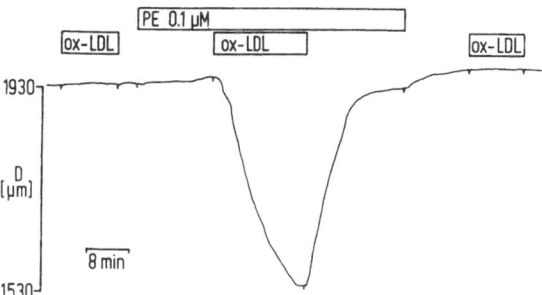

Fig. 6. Diameter (D) recording of a rabbit femoral artery segment, intraluminally perfused with ox-LDL (80 μg/ml), either in the absence or in the presence of phenylephrine (*PE*) in subthreshold concentration (0.1 μM). ox-LDL caused no vasoconstriction on its own, but substantial vasoconstriction in the presence of phenylephrine. *Bars* indicate the presence of ox-LDL or phenylephrine

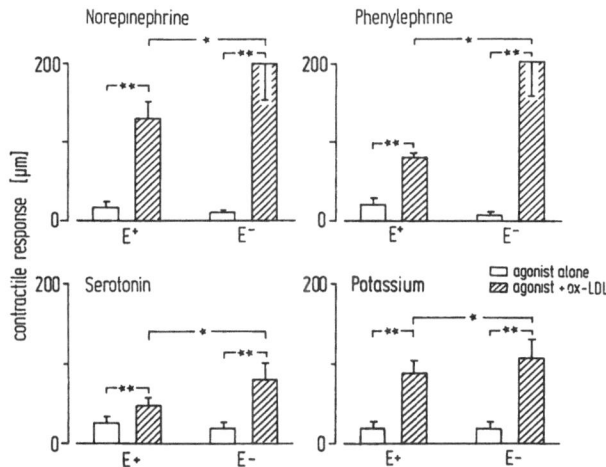

Fig. 7. Perfusion with ox-LDL enhances contractile responses to various agonists. Agonists were administered to the organ bath at threshold concentrations which evoked only marginal vasoconstrictions in segments with (E^+) and without (E^-) endothelium. NE, norepinephrine [1–3 nM, 17 ± 8 μm (E^+), 12 ± 5 μm (E^-)]; PE, phenylephrine [10–30 nM, 7 ± 5 μm (E^+), 21 ± 9 μm (E^-)]; 5-HT, serotonin [10–50 nM, 26 ± 7 μm (E^+), 20 ± 7 μm (E^-); K+, potassium [20–30 mM, 20 ± 8 μm (E^+), 19 ± 8 μm (E^-)]. ox-LDL (80 mg/ml) was perfused intraluminally, and caused only minimal vasoconstrictions [9 ± 2 μm (E^+), 17 ± 4 μm (E^-)] in the absence of agonists. Contractile responses (μm) are expressed as means $\pm$ SEM. Each panel shows the enhancement of contractile responses to one agonist by ox-LDL in endothelium-intact and endothelium-denuded segments. $^*P < 0.05$, $^{**}P < 0.01$; $n = 7$ for each group

phenylephrine (PE), serotonin (5-HT) and potassium (K^+), all administered at threshold concentrations, were significantly augmented in the presence of

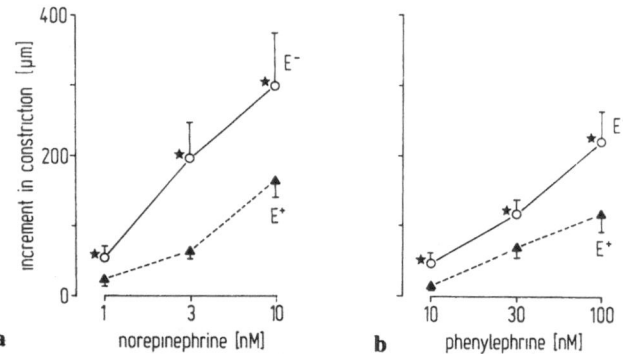

Fig. 8a,b. Increase in potency of norepinephrine (*NE*) and phenylephrine (*PE*) by ox-LDL in rabbit femoral arteries. NE (0.3–10 nM) and PE (0.3–100 nM) were administered in cumulative doses to the intraluminal perfusion, with and without ox-LDL (80 µg/ml). *Triangles* represent increments of contractile responses (means ± SEM) to **a** Ne and **b** PE when ox-LDL was additionally administered to the intraluminal perfusion of endothelium-intact (E$^+$) segments; *circles* represent increments of contractile responses by ox-LDL in endothelium-denuded (E$^-$) segments. In segments without endothelium, the increments of contractile responses were significantly greater than in segments with endothelium. ox-LDL alone caused no vasoconstriction. $^*P < 0.05$

ox-LDL (80 µg/ml). Native LDL did not enhance the contractile responses to these agonists. For each agonist, the enhancement was significantly greater in endothelium-denuded than in endothelium-intact segments. In the case of NE and PE, cumulative concentration-response relationships were obtained with and without ox-LDL. As shown in Fig. 8, ox-LDL increased the potency of NE and PE in endothelium-intact and endothelium-denuded segments. Again, the enhancement of the contractile response was significantly greater in endothelium-denuded than in endothelium-intact segments.

In additional experiments, ox-LDL derived by prior incubation with activated macrophages also enhanced contractile responses to norepinephrine (data not shown).

Effects of Ca^{2+}-Antagonists and Indomethacin upon ox-LDL Evoked Vasoconstrictions

To investigate whether ox-LDL evoked vasoconstriction is linked to transmembraneous Ca^{2+}-influx, we preincubated segments with three structurally different types of Ca^{2+}-channel blockers: diltiazem, verapamil and nitrendipine. As shown in Fig. 9. ox-LDL induced vasoconstriction was nearly equally suppressed by diltiazem, verapamil and nitrendipine in the presence of norepinephrine. These inhibitory effects were more pronounced in endothelium-denuded than in endothelium-intact segments (Fig. 9, Table 1). However, the Ca^{2+}antagonist-induced suppression was different for the specific agonists used (Table 1).

Fig. 9. Suppression of ox-LDL (80 μg/ml) evoked contractile responses by the Ca^{2+}-antagonists verapamil (1 μm), diltiazem (10 μM), and nitrendipine (1 μm) in the presence of norepinephrine ($n = 6$). Suppression is expressed in percent of the contractile response. The suppression by diltiazem, verapamil, and nitrendipine was more effective in endothelium-denuded (E^-) than in endothelium-intact (E^+) *segments*. *$P < 0.05$

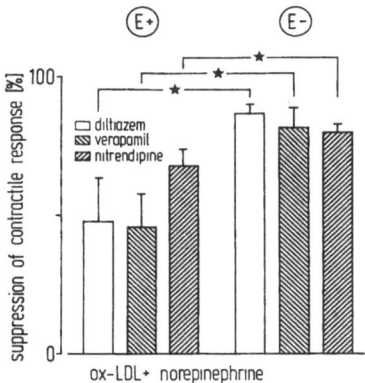

Results similar to those obtained in the presence of norepinephrine were also found in the absence of contractile agonists. Vasoconstriction induced by ox-LDL alone was reduced by verapamil by 52 ± 4% (diltiazem: 65 ± 6%) in endothelium-intact segments, and by 71 ± 5% (diltiazem: 77 ± 5%) in endothelium-denuded segments ($n = 6$).

When the segments were perfused with the cyclooxygenase inhibitor indomethacin (10 μM), no change in the potentiation of contractile responses to ox-LDL was detected (data not shown).

Discussion

The data presented in this study indicate that both n-LDL and ox-LDL inactivate EDRF after its release from endothelial cells, and that not native LDL, but its oxidized derivatives enhance contractile responses of the isolated rabbit femoral artery to various agonists. Native LDL by itself, neither induced nor enhanced (in the presence of various contractile agonists) vasomotor responses. The formation of EDRF, which is most likely NO [32], remained unaffected after 1 h incubation with n-LDL and ox-LDL, both in cultured endothelial cells and in isolated arterial segments.

There is strong evidence suggesting that both native and oxidized LDL are present in the arterial wall in hypercholesterolemia [15–18]. Thus, it is conceivable that the inactivation of EDRF by n-LDL and ox-LDL contributes to the impairment of endothelial function in hypercholesterolemia, as it has been observed in numerous studies [2–5,8,33]. Also, the markedly reduced responsiveness of atherosclerotic arteries to EDRF/NO-superfusion [34] can be explained by this direct inactivating effect of n-LDL and ox-LDL accumulating in fatty streaks and atherosclerotic plaques. Furthermore, the increased responsiveness to vasoconstricting agents observed in arteriosclerotic human [2] and animal arteries [2,4,7–10], and their increased disposition for vasospasm might be explained by the potentiating effect of ox-LDL on vascular smooth muscle contractions.

However, our observation that neither a short exposure (1 h) to n-LDL, nor to ox-LDL attenuates the formation of EDRF, is in contrast to conclusions drawn by others. In studies performed with arterial strip preparations. Andrews et al. [14] and Tomita et al. [35] found impairment of endothelium-dependent vasodilation after less than 30 min incubation with 'native' LDL. Henry et al. observed impairment of endothelial function with ox-LDL (2 h incubation) [36], and Vedernikov et al. [37] found, dependent on the route of LDL-application and on the origin of the lipoproteins, both endothelium-dependent vasorelaxation induced by LDL, and impairment of acetylcholine-induced vasodilation (after 30 min incubation with LDL).

Different techniques applied for investigation of effects of the lipoproteins on vasomotion could explain the various findings. In our experimental approach with intact segments, endothelium-mediated vasodilation induced by acetylcholine was not impaired after incubation with n-LDL or ox-LDL. However, incubation of arterial ring preparations with n-LDL inhibited such relaxation. One can speculate that arterial ring or strip preparations are easier permeated by lipoproteins than intact segments. This rapid permeation would render the direct EDRF-inactivating effect of n-LDL and ox-LDL, and thus explain why endothelium-dependent vasodilation was attenuated after relatively short incubation with lipoproteins in ring and strip preparations [14,35–37], but not in intact segments.

Furthermore, the comparison of studies undertaken with n-LDL or ox-LDL is associated with major diffculties. Absence of adequate measures against lipid peroxidation favors the development of oxidatively modified LDL. Hence, without detection of the oxidative state of the LDL preparation [35,37], the observed effects cannot be unequivocally attributed to native LDL.

Oxidized LDL is not clearly defined, and during the lipid peroxidation process, a variety of partly unstable, biologically active substances is formed [38]. ox-LDL can be cytotoxic to endothelial cells [13,39,40], but its composition can differ depending on the plasma origin and the oxidative conditions [41]. Because of this problem, we used lipoproteins prepared using a standardized protocol throughout the whole series of experiments, in all the systems studied. A biological test for our standardization was that oxidized LDL had the capacity to potentiate vasoconstriction as reported in a previous study [31].

The fact that EDRF-formation was not attenuated by 1 h incubation with the potentially cytotoxic ox-LDL can be explained by the rather short incubation time, and this must be clearly differentiated from chronic effects. Thus, we may have missed cytotoxic effects of ox-LDL occuring with more prolonged exposure.

The molecular mechanism responsible for inactivation of EDRF/NO by LDL is unknown. A possible desensitization of the vascular smooth muscle to EDRF by the lipoproteins seems unlikely. When n-LDL was added in the bioassay experiments *during* continuous EDRF-perfusion of the detector segment, vasodilations were suppressed immediately. Also, the suppression was immediately reversible by lipoprotein washout. This suggests direct inactivation of EDRF by lipoproteins, rather than an effect on the target organ smooth muscle.

The highly hydrophobic core of the LDL particle may act as a sink for EDRF/NO, which is approximately 8 times as soluble in hydrophobic than in hydrophilic media [42]. The NO radical could be consumed by reactions with hydrocarbonic radicals inside the LDL particle. However, the exact mechanism of NO-inactivation by LDL remains to be clarified.

The mechanism of the ox-LDL-induced enhancement of contractile responses also remains to be determined. The enhancing effect was observed in endothelium-intact segments, and even more pronounced in endothelium-denuded segments, providing evidence that the site of action is at the smooth muscle itself, rather than at the endothelium. Further evidence that the endothelium was not involved in the enhancing mechanism is based on an undisturbed production of EDRF even after 2 h incubation of the segments with ox-LDL in concentrations which already markedly augmented agonist-induced vasoconstriction. This certainly does not rule out that ox-LDL can be cytotoxic: but if so, it would predominantly act on the smooth muscle cells.

The fact that the enhancing effect of ox-LDL was more pronounced in endothelium-denuded segments might be due to the loss of the basal release of EDRF [43]. This enhancement of contractile responses following endothelium-removal has already been described in earlier studies [19,43–45].

The suppression of contractile responses by the Ca^{2+}-antagonists diltiazem, verapamil and nitrendipine was also more effective in endothelium-denuded segments. Similar findings were made with different nitrovasodilators, which were also more effective in endothelium-denuded than in endothelium-intact arterial segments [46,47].

The suppressor effect of Ca^{2+}-antagonists on vasoconstriction elicited by ox-LDL, both in the presence and in the absence of contractile agonists, suggests that ox-LDL predominantly induces an increased transmembraneous Ca^{2+}-influx. However, an additional release of Ca^{2+} from intracellular stores cannot be excluded.

The possibility that ox-LDL sensitizes the contractile apparatus to Ca^{2+} is less likely. The enhancing effect started after a few minutes, and washout also reversed the effect within 10 min. It is inconceivable that ox-LDL enters and leaves the smooth muscle cells within this short period of time.

Since the enhancing effect is observed with various receptor binding agonists, and also with receptor-independent K^+-depolarization, ox-LDL seems to act distally to the receptor- coupled signal transduction cascade. One can speculate that, via stimulation of phosphatidylinositol metabolism, ox-LDL modulates the voltage-gated Ca^{2+}-channels, and thus the transmembraneous Ca^{2+}-influxes in the plasma membrane. However, further investigations are needed to clarify the molecular mechanism.

In conclusion, our findings demonstrate that both native and oxidized LDL directly inactivate EDRF, and that oxidized LDL enhances agonist-induced vasoconstriction by direct interaction with vascular smooth muscle. Formation of EDRF was not attenuated after short term exposure of endothelial cells and intact segments to the lipoproteins. The EDRF-inactivating effect of both lipoproteins, and the vasoconstriction-potentiating effect of ox-LDL, may be of particular pathophysiological relevance in regions with lipoprotein accumulation in the vessel wall, such as fatty streaks and atherosclerotic plaques, and may favor the initiation of inappropriate vasoconstriction.

References

1. Steinberg D, Parthasarathy S, Carew TE, Khoo JC, Witztum JL (1989) Beyond cholesterol: Modifications of low-density lipoprotein that increase its atherogenicity. N Engl J Med 320: 915–924
2. Bossaller C, Habib GB, Yamamoto H, Williams C, Wells S, Henry PD (1987) Impaired muscarinic endothelium-dependent relaxation and cyclic guanosine 5'-monophosphate formation in atherosclerotic human coronary artery and rabbit aorta. J Clin Invest 79: 170–174
3. Osborne JA, Lento PH, Siegfried MR, Stahl GL, Fusman B, Lefer AM (1989) Cardiovascular effects of acute hypercholesterolemia in rabbits. J Clin Invest 83: 465–473
4. Hof RP, Hof A (1988) Vasoconstrictor and vasodilator effects in normal and atherosclerotic conscious rabbits. Br J Pharmacol 95: 1075–1080
5. Verbeuren T, Jordaens F, Zonnekeyn L, Van Hove C, Coene M, Herman A (1986) 1. Endothelium-dependent and endothelium-independent contractions and relaxations in isolated arteries of control and hypercholesterolemic rabbits. Circ Res 58: 552–564
6. Wines PA, Schmitz JM, Pfister SL, Clubb FJ, Buja LM, Willerson JT, Campbell WB (1989) Augmented vasoconstrictor responses to serotonin precede development of atherosclerosis in aorta of WHHL rabbit. Arteriosclerosis 9: 195–202
7. Rosendorf C, Hoffman JIE, Verrier ED, Rouleau J, Boerboom LE (1981) Cholesterol potentiates the coronary artery response to norepinephrine in anesthetized and conscious dogs. Circ Res 48: 320–329
8. Tomoike H, Egashira K, Yamamoto Y, Nakamura M (1989) Enhanced responsiveness of smooth muscle, impaired endothelium-dependent relaxation and the genesis of coronary spasm. Am J Cardiol 63: 33E–39E
9. Henry PD, Yokoyama M (1980) Supersensitivity of atherosclerotic rabbit aorta to ergonovine. Mediation by a serotonergic mechanism. J Clin Invest 66: 306–313
10. Heistad DD, Armstrong ML, Marcus ML, Piegors DJ, Mark AL (1984) Augmented responses to vasoconstrictor stimuli in hypercholesterolemic and atherosclerotic monkeys. Circ Res 54: 711-718
11. Holland JA, Pritchard KA, Rogers NJ, Stemerman MB. Perturbation of cultured human endothelial cells by atherogenic levels of low density lipoprotein (1988) Am J Phathol 132: 474–478
12. Cathcart MK, Morel DW, Chisolm GM. Monocyte and neutrophils oxidize low density lipoprotein making it cytotoxic (1985) J Leukocyte Biol 38: 341–350
13. Hennig B, Chow CK (1988) Lipid peroxidation and endothelial cell injury: implications in atherosclerosis. Free Radical Biol Med 4: 99–106
14. Andrews HE, Bruckdorfer KR, Dunn RC, Jacobs M (1987) Low-density lipoproteins inhibit endothelium-dependent relaxation in rabbit aorta. Nature 327: 237–239
15. Hoff HF, Morton RE (1985) Lipoproteins containing apoB extracted from human aortas. Ann NY Acad Sci 454: 183–194
16. Daugherty A, Zweifel BS, Sobel BE, Schonfeld G (1988) Isolation of low density lipoprotein from atherosclerotic vascular tissue of Watanabe heritable hyperlipidemic rabbits. Arteriosclerosis 8: 768–777

17. Palinski W, Rosenfeld ME, Ylä-Herttuala S, Gurtner GC, Socher SS, Butler SW, Parthasarathy S, Carew TE, Steinberg D, Witzum JL (1989) Low density lipoprotein undergoes oxidative modification in vivo. Proc Natl Acad Sci USA 86: 1372–1376
18. Ylä-Herttuala S, Palinsky W, Rosenfeld ME, Parthasarathy S, Carew TE, Butler S, Witzum JL, Steinberg D (1989) Evidence for the presence of oxidatively modified low density lipoprotein in atherosclerotic lesions of rabbit and man. J Clin Invest 84: 1086–1095
19. Bassenge E, Busse R (1988) Endothelial modulation of coronary tone. Prog Cardiovasc Dis 30: 349–380
20. Edelstein C, Scanu AM (1986) Precautionary measures for collecting blood destinated for lipoprotein isolation. In: Segrest JP, Albers JJ (eds) Methods in enzymology. Academic Press, New York, pp 151–155
21. Havel RJ, Eder HA, Bragdon JH (1955) The distribution and chemical composition of ultracentrifugally separated lipoproteins in human serum. J Clin Invest 34: 1345–1353
22. Bradford M (1975) A rapid and sensitive method for the quantitation of microgram quantities of protein utilizing the principle of protein dye binding. Anal Biochem 72: 248–254
23. Esterbauer H, Striegel G, Puhl H, Rotheneder M (1989) Continuous monitoring of in vitro oxidation of human low density lipoprotein. Free Radical Res Comm 6: 67–75
24. Steinbrecher UP, Witztum JL, Parthasarathy S, Steinberg D (1987) Decrease in reactive amino groups during oxidation or endothelial cell modification of LDL. Correlation with changes in receptor-mediated catabolism. Arteriosclerosis 7: 135–143
25. Laemmli UK (1970) Cleavage of structural proteins during the assembly of the head of bacteriophage T4. Nature 227: 680–685
26. Lückhoff A, Busse R, Winter I, Bassenge E (1987) Characterization of vascular relaxant factor released from cultured endothelial cells. Hypertension 9: 295–303
27. Förstermann U, Goppelt-Strübe M, Frolich JC, Busse R (1986) Inhibitors of acyl-coenzyme A: Lysolecithin acyltranferase activates the production of endothelium-derived vascular relaxing factor. J Pharmacol Exp Ther 238: 352–359
28. Mülsch A, Böhme E, Busse R (1987) Stimulation of soluble guanylate cyclase by endothelium-derived relaxing factor from cultured endothelial cells. Eur J Pharmacol 135: 247–250
29. Lückhoff A, Mülsch A, Busse R (to be published) cAMP attenuates autacoid release from endothelial cells: relation to internal calcium. Am J Physiol
30. Busse R, Pohl U, Kellner C, Klemm U (1983) Endothelial cells are involved in the vasodilatory response to hypoxia. Pflügers Arch 397: 78–80
31. Galle J, Bassenge E, Busse R (1990) Oxidized low density lipoproteins potentiate vasoconstrictions to various contractile agonists by direct interaction with vascular smooth muscle. Circ Res 66: 1287–1293
32. Palmer RMJ, Ferrige AG, Moncada S (1987) Nitric oxide release accounts for the biological activity of endothelium-derived relaxing factor. Nature 327:524–526
33. Cox DA, Vita JA, Treasure CB, Fish RD, Alexander RW, Ganz P, Selwyn AP (1989) Atherosclerosis impairs flow-mediated dilation of coronary arteries in humans. Circulation 80: 458–465
34. Verbeuren TJ, Jordaens FH, VanHove CE, VanHoydonck AE, Herman AG (to be published) Release and vascular activity of the endothelium-derived relaxant factor in atherosclerotic rabbit aorta. Eur J Pharmacol
35. Tomita T, Ezaki M, Miwa M, Nakamura K, Inoue Y (1990) Rapid and reversible

inhibition by low density lipoprotein of the endothelium-dependent relaxation to hemostatic substances in porcine coronary arteries. Heat and acid labile factors in low density lipoprotein mediate the inhibition. Circ Res 66: 18–27

36. Kugiyama K, Bucay M, Morrisett JD, Roberts R, Henry PD (1989) Oxidized LDL impairs endothelium-dependent arterial relaxation. Circulation 80 (Suppl 2): 279

37. Vedernikov Y, Lankin V, Tikhaze A, Vikhert A (1988) Lipoproteins as factors in vessel tone and reactivity modulation. Basic Res Cardiol 83: 590–596

38. Esterbauer H, Jürgens G, Quehenberger O, Koller E (1987) Autoxidation of human low density lipoprotein: Loss of polyunsaturated fatty acids and vitamin E and generation of aldehydes. J Lipid Res 28: 495–509

39. Morel DW, Hessler JR, Chisolm GM (1983) Low density cytotoxicity induced by free radical peroxidation of lipid. J Lipid Res 24: 1070–1076

40. Hessler JR, Robertson AL, Chisolm JR, Chisolm GM (1979) LDL induced cytotoxicity and its inhibition by HDL in human vascular smooth muscle and endothelial cells in culture. Ateriosclerosis 32: 213–229

41. Esterbauer H, Rotheneder M, Striegel G, Waeg G, Ashy A, Sattler W, Jürgens G (1989) Vitamin E and other lipophilic antioxidants protect LDL against oxidation. Fat Sci Technol 8: 316–324

42. Link WF (1965) Solubilities of inorganic and metal-organic compounds vol 2, 4th edn. American Chemical Society, Wasington DC, p 792

43. Griffith TM, Henderson AH, Edwards DH, Lewis MJ (1984) Isolated perfused rabbit coronary artery and aortic strip preparations: the role of endothelium-derived relaxant factor. J Physiol (Lond) 351: 13–24

44. Cohen RA, Zitnay KM, Weisbrod RM, Tesfamariam B (1988) Influence of the endothelium on tone and the response of isolated pig coronary artery to norepinephrine. J Pharmacol Exp Ther 244: 550–555

45. Chiba S, Tsukada M (1984) Potentiation of KCl-induced vasoconstriction by saponin treatment in isolated canine mesenteric arteries. Jpn J Pharmacol 36: 535–537

46. Pohl U, Busse R (1987) Endothelium-derived relaxant factor inhibits the effect of nitrocompounds in isolated arteries. Am J Physiol 252: H307–H313

47. Shirasaki Y, Su C (1985) Endothelium removal augments vasodilation by sodium nitroprusside and sodium nitrite. Eur J Pharmacol 114: 93–96

CHAPTER 5

Pathophysiology of Ischemic Heart Disease with Special Reference to Coronary Artery Spasm

M. Nakamura[1]

Summary. Traditionally, ischemic heart disease has been considered the sequence of fixed atherosclerotic obstruction of the large epicardial coronary arteries. However, clinical hemodynamic, angiographic, scintigraphic and metabolic and animal studies have provided independent evidence for an increase in vasomotor tone, especially coronary spasm, as an important cause of various stages of ischemic heart disease. Factors that initiate or sustain the syndrome related to coronary spasm have not been completely defined, and a porcine model of coronary spasm was prepared in our laboratory.

The hypothesis of coronary spasm as a cause of angina pectoris was proposed during the 19th century. In the 1920's, the close association between angina and atherosclerotic coronary narrowing was confirmed by pathological studies and coronary angiographic support was obtained in 1960s and 1970s, which neglected the presence of coronary spasm.

Beginning in the 1930s, there were reports of spontaneous angina not associated with increased myocardial oxygen demand. In 1959, Prinzmetal et al. summarized these cases, and reported a variant form of angina pectoris, ignoring the myocardial oxygen demand-supply imbalance theory. They proposed that increased coronary artery tonus was the cause of this variant angina. During the 1970s, Maseri and other investigators clearly demonstrated that coronary spasm is a major cause of variant angina.

Coronary spasm is now accepted as a major cause of variant angina, a significant proportion of rest angina and rest and effort angina, some cases of effort angina and postmyocardial infarctional angina, and possibly acute ischemia related death including acute myocardial infarction. However the link between coronary spasm, progression of organic stenosis, and thrombotic occlusion of coronary arteries is missing.

The pathophysiology of acute myocardial infraction is also uncertain. Recent clinical and pathophysiological studies have indicated that coronary-occlusive thrombi occur frequently in the early stage of transmural infarct, but infrequently with subendocardial infarct and sudden ischemic death. Mechanisms and factors which initiate coronary thrombosis are unknown. Falk deduced from pathological studies that rupture of the atheromatous plaque surface was always accompanied by hemorrhage into the plaque, but occlusive thorombosis was rare unless the rupture was associated with a stenosis greater than 75%. Thrombi in arteries are frequently located on the surface of

---

[1] Research Institute of Angiocardiology and Cardiovascular Clinic, Faculty of Medicine, Kyushu University, 3-1-1 Maidashi, Higashi-ku, Fukuoka, 812 Japan

atherosclerotic plaques, and not necessarily in the arteries related to myocardial infarction.

The antithesis to thrombosis as a cause of acute myocardial infarction is intramural hemorrhage in the atheroma as the primary event causing coronary occlusion. Clinical evidence supporting coronary spasm as the pathogenesis of acute myocardial infarct has been reported. The recent hypothesis that acute myocardial infarction occurs as a direct result of coronary spasm has not been fully proven. Thus, various studies on the development of suitable animal models of coronary spasm have been performed.

An animal model of coronary spasm should be helpful in clarifying mechanisms and roles of spasm in various stages of ischemic heart disease, including progression of organic stenosis and possibly acute coronary occlusion.

Gensini et al. were the first to demonstrate diffuse coronary spasm by giving pitressin to a normal dog, in 1962. In 1983, we developed a porcine model of coronary spasm in mildly atherosclerotic miniature swine, by injecting histamine and serotonin. We have recently improved the animal model of coronary spasm in moderate coronary atherosclerosis by giving serotonin, ergonovine, and hyperventilation, similar to what occurs in patients with variant angina.

In isolated coronary artery segments from atherosclerotic miniature swine, we found that a decrease in endothelium-dependent relaxing function and abnormal hypercontraction of the media of atherosclerotic coronary arteries. These events may result from an increased number of receptors for agonists, and/or augumentation of signal transduction, but not by increased calcium sensitivity of contractile proteins in the medial smooth muscle cells.

I will discuss about the following; a) the role of preexisting coronary atherosclerosis, b) neural factors, such as the adrenergic and parasympathetic nervous systems, c) autacoids, such as histamine and serotonin in the development of coronary spasm, and d) variations in in vitro, in vivo, and inter-species evidence.

Introduction

Traditionally, ischemic heart disease has been considered the consequence of fixed atherosclerotic obstruction of the coronary artery. However, clinical hemodynamic, angiographic, scintigraphic, and metabolic studies, as well as animal studies have provided independent evidence for the role of coronary arterial spasm in the etiology of various stages of ischemic heart disease. Factors that initiate or sustain syndromes related to coronary spasm have not been completely defined, and a suitable animal model was needed. In this review article, attention will be directed to the pathophysiology of ischemic heart disease, with special reference to coronary spasm.

History of the Concept of Coronary Spasm and Evidence for its Role in Angina Pectoris

The hypothesis of transient coronary spasm as a causative factor for angina pectoris was proposed during the nineteenth century [1]. Osler [2] stated that "spasm or narrowing of a coronary artery or even one branch, may so modify the action of a section of the heart that it works with disturbed tension and there are stretching and strain sufficient to arouse painful sensations. . . . I do not know of any better explanation of anginal pain".

The antithesis to coronary spasm appeared in the early 1920s. Keefer and Resnik emphasized that the atherosclerotic coronary artery could not constrict, and the role of coronary vasomotion in angina pectoris was ignored [3]. A close association between atherosclerotic fixed narrowing of the coronary arteries and angina was noted in pathological studies and support was given to the notion that imbalance between myocardial oxygen demand and supply is a major cause of angina [4,5], and coronary angiographic evidence for this was subsequently obtained [6]. Later, the failure to find coronary spasm with any frequency in numerous coronary angiograms performed over the previous two decades, focused attention only on organic stenosis as underlying mechanism for angina pectoris, probably because in many laboratories nitroglycerin was routinely administered prior to coronary angiography. This protocol tended to focus attention away from the possible role of coronary vasomotion. Coronary spasm can be missed if not actually sought, and at times may possibly be confused with organic atherosclerotic obstruction. Very little evidence, if any, of coronary spasm was obtained at autopsy.

From 1931, reports were made of attacks of spontaneous angina in association with transient ST elevation, and in some instances, angina provoked by ergonovine [1]. In 1959, Printzmetal et al. [7] described a variant form of angina pectoris which differed from classic effort angina. The temporary increased tonus (spasm) of the epicardial large coronary arteries was suggested to be a cause of angina, without an increase in myocardial oxygen demand. They also noted that the area of the myocardium giving rise to ST segment elevation during anginal pain is often the site of future myocardial infarction. They reported a high incidence of sudden death and myocardial infarction in patients with variant angina. The absence of an increase in heart rate and systolic blood pressure before anginal attacks was noted by several investigators, thereby excluding an increased metabolic demand as a cause of variant angina attack [8,9]. Data presented at the conference of the National Heart and Lung Institute, USA, in 1971 indicated that a common cause of precipitation of angina pectoris, even without exercise, is an increase in factors responsible for increasing myocardial oxygen demand. However, little was said about the role of coronary vasomotion (10). In the 1970s, Maseri et al. [11–13] demonstrated that anginal pain and changes in blood pressure were preceded by a decreased oxygen content of the coronary sinus. This suggested a primary decrease in coronary flow. A transient complete occlusion of one of the major coronary arteries, and a perfusion defect on ^{201}Tl myocardial scintigraphy were found during the anginal attack. These findings ignored the possibility of catheter-induced spasm.

The importance of coronary spasm as a pathogenic factor in ischemic heart disease continues to be of much interest in various specialties. Coronary spasm was noted to occur more frequently in patients with mild atherosclerosis, even in coronary arteries with normal appearance, and in particularly in Japanese people [14–17]. In our experience, coronary artery vasospasm was a major factor in more than one third of patients with angina pectoris admitted to our clinic. Calcium channel blockers, such as diltiazem and nifedipine completely prevented anginal attacks in about 90% of patients with variant angina, when given in relatively smaller doses compared to those prescribed for patients with a non-spastic angina [15]. Hiramori reported that the rate of incidence of coronary spasm gradually decreased with progression of the degree of organic coronary stenosis [16]. We also reported that the risk of sudden death and malignant arrhythmia in patients with vasospastic angina did not depend on the presence of significant stenosis. However, the risk of myocardial infarction was significantly higher in patients with a fixed stenosis of 90% or greater, than in patients with stenosis less than 90% or normal coronary arteries [17]. Other workers also reported that exercise testing may trigger coronary spasm, followed by acute myocardial ischemia [18,19]. The occurrence of spasm at different sites in the same or different coronary arteries in the same patient and at different times may explain the variability of the clinical manifestations observed in a certain number of patients with coronary atherosclerosis [13].

Thus, coronary spasm is one of the major causes of variant angina, a significant percentage of rest angina, and rest and effort angina [20], some cases of effort angina [18,19], and post myocardial infarction angina [21], and possibly acute ischemic death, including acute myocardial infarction [22,23]. However, we still do not know the nature of the link between coronary spasm, progression of organic stenosis, and thrombotic occlusion in the acute obstruction of coronary arteries. Marzilli et al. [24] claimed that coronary spasm can be an antecedent factor leading to later development of fixed atherosclerotic coronary arterial obstruction. However, this hypothesis has not gained wide support [25].

At our research laboratory, coronary spasm was induced by administering histamine and serotonin to atherosclerotic miniature swine [26]. Severe coronary spasm can induce intramural hemorrhage at the site of spasm, followed by sudden progression of organic stenosis [27]. Details of our swine model will be described later.

The Role of Coronary Spasm in the Development of Acute Myocardial Infarction

The pathophysiology of acute myocardial infarction is also open to question. There appears to be general agreement that coronary thrombi-producing occlusion do not frequently occur in patients who die suddenly and in those with subendocordial infarcts, although coronary atherosclerotic narrowing is usually present. However, extensive transmural myocardial infarction is commonly associated with total coronary occlusion observed angiographically during the early hours following the acute event [28]. In addition, evidence of coronary thrombotic occlusion in acute myocardial infarction was obtained when much attention was directed to the widespread trial of thrombolytic therapy for acute myocardial infarction. However, these studies did not provide information on the natural history of thrombus formation, or the precise mechanisms involved in the initiation of heart attack, and the relationship between thrombi and spasm. Clinically, an association between coronary spasm and thrombosis has been suggested [22,23,29–31]. Maseri et al. emphasized that coronary spasm can occur at the onset of acute myocardial infarction in patients with previous unstable angina [22]. However, several other studies demonstrated that the incidence of coronary spasm in acute myocardial infarction is relatively infrequent [32–34].

Lesions in cases of coronary atherosclerosis, a commonly accepted cause of coronary occlusion include these features: (1) progressive luminal narrowing, (2) thrombosis, (3) hemorrhage into the atheromatous plaque, and (4) rupture of the atheromatous plaque [35]. In cases of atherosclerosis seen in Watanabe Heritable Hyperlipidemic rabbits in which a state of hypercholesterolemia had been maintained for one year, severe organic coronary stenosis and myocardial necrosis occurred, but there was no thrombus formation [36].

Recent pathological studies have emphasized that thrombus formation is frequently associated with rupture of the atheromatous plaque in the presence of severe stenosis. Falk described that rupture of the plaque surface was always accompanied by hemorrhage into the plaque, but occlusive thrombosis was rare unless the rupture was associated with a stenosis greater than 75% [37]. Thrombi in arteries are frequently present over the atherosclerotic plaque, but not always in the arteries related to myocardial infarction [38].

The antithesis to thrombus formation as a cause of acute myocardial infarction is hemorrhage into the atheromatous plaque as the primary event causing acute coronary occlusion [39,40]. Intramural hemorrhage into the atheromatous plaque may damage the endothelium by increasing the volume and pressure within the plaque, and thus enhance thrombosis. Singh [41] reported that intimal hemorrhages are the likely explanation for the intimal encroachment and the episodic nature of progression of coronary artery stenosis, as shown by repeated coronary arteriographic studies performed for coronary artery disease. He emphasized that coronary atherosclerosis does not progress gradually in a linear fashion, and local anatomical factors appear to play a dominant role in the natural history of ischemic heart disease. Davies and Thomas proposed two different processes in plaque hemorrhage [42]. The first is that hemorrhage from neovascularized vessels is ubiquitous in a large lipid-rich plaque. The second is hemorrhage through a fissure in the plaque, which may be related to occlusive thrombi.

Clinical observations suporting coronary spasm in the pathogenesis of acute myocardial infarction are as follows: (1) Acute myocardial infarction frequently occurs in patients with vasospastic angina, as described above [7,17,22,31,43]. Myocardial infarction frequently occurs in the area supplied by the vessels which were shown to undergo spasm in patients with variant angina. (2) The frequency of coronary spasm in patients with unstable angina is significant [12,31,44–47]. Thus, there is circumstantial evidence linking coronary spasm and platelet activation and/or thrombosis. (3) It was reported that in some patients with no significant coronary stenosis, that acute myocardial infarction occurs after induction of coronary spasm induced by the administration of ergonovine (48,49). Also, nonocclusive coronary thrombus associated with provoked spasm led to persistent coronary occlusion and myocardial infarction. (4) Bertrand et al. [50] gave ergometrine to asymptomatic patients less than 6 weeks after acute myocardial infarction, and noted that coronary artery spasm occurred in 20% of these patients. (5) We have reported that spontaneous or provoked coronary spasm occurred in some patients with post infarctional angina [21]. In these patients, coronary spasm was usually of the same site as the lesion which caused infarction. (6) Maseri et al. reported that in one patient with spasm at the onset of infarction, postmortem examination revealed fresh thrombi at the spastic area [22]. (7) Sudden withdrawal from a work environment associated with exposure to a high concentration of nitrates in a person employed in the munitions industry was followed by acute myocardial infarction, angina pectoris and sudden death, in the absence of obstructive coronary stenosis. In some of these patients, angiography revealed evidence of coronary spasm which was relieved by nitroglycerine [51]. We also found that when calcium antagonists were given to patients with variant angina, sudden withdrawal of calcium blockade sometimes caused an increase in the frequency of angina, sudden death, or acute myocardial infarction [14]. However, the hypothesis that acute myocardial infarction develops as a result of coronary spasm, awaits further support.

The first clinical sign of ischemic heart disease can be sudden death and/or acute myocardial infarction, and their prevention is still an unsolved problem since we do not know all the mechanisms involved in so-called "heart attack".

Definition of Coronary Spasm

In discussing the roles of coronary spasm in the development of ischemic heart disease, the term "spasm" must be clearly defined. Maseri and Chierchia proposed a broad definition of coronary spasm as "inappropriate, active constriction, focal or diffuse, of one or more large conductive coronary arteries" [13]. This definition does not exclude constriction of coronary artery branches which is not sufficiently severe to cause acute myocardial ischemia. Another arbitrary definition is "the transient stenosis of an epicardial coronary artery of sufficient degree to produce myocardial ischemia that is promptly reversed by nitroglycerin in the absence of evidence for increased myocardial oxygen demand" [43]. This definition dose not set limits on the degree of fixed organic coronary artery stenosis before spasm occurs, and fits the events in patients with severe organic stenosis. However, MacAlpine proposed that acute luminal narrowing occurring at stenotic sites as the result of "nomal" vasomotion, and indicated that even modest mural thickening may act as a lever in translating physiologic degrees of medial smooth muscle shortening into critical luminal obstructions [52]. Ergonovine will produce about a 20% narrowing of the diameter of the coronary artery in patients with no coronary spasm. Coronary spasm has been defined by other workers as a transient narrowing of 85% – 100% of the diameter of an epicardial coronary artery [53]. In present day medicine, hypercontraction of the resistance vessels in the coronary circulation is usually excluded from the definition of coronary spasm, because of difficulties in defining the abnormalities. Leucotrienes, 5-lipoxygenase metabolites of arachidonic acid [54], and endothelin, a potent vasoconstrictive peptide discovered in the culture medium of endothelial cells, did not provoke spasm of the epicardial coronary arteries, but did cause severe hypercontraction of small coronary arteries resulting, in myocardial ischemia [55]. At present, coronary arteriography is the only available tool which can effectively demonstrate the occurrence of coronary spasm. A definition of coronary spasm according to tension studies of isolated coronary artery has not yet been proposed, and quantitative or qualitative in vitro differences between spasm and hypercontraction after vasoactive substances remain undefined. Most cardiologists consider coronary spasm as a state when the diameter of a large coronary artery with less than 50% stenosis becomes totally or subtotally obstructed, in association with acute myocardial ischemia, which is relieved by nitroglycerin. Extensive clinical and postmortem research has failed to detect mechanisms or triggering stimuli for heart attacks.

Animal Models of Coronary Spasm

Establishment of an animal model of coronary spasm would be expected to clarify the mechanisms of spasm and possibly of related ischemic heart diseases, including the relationship between organic and dynamic obstruction, as well as between spasm and thrombosis of coronary arteries.

History of the Development of Experimental Coronary Spasm

Gensini et al. first demonstrate diffuse coronary spasm angiographically in normal dogs, by administration of pitressin. This was associate with a decrease in coronary venous occluded pressure, suggesting a decrease in coronary flow [56]. Perez et al. found reversible, complete coronary occlusion associated with reduced coronary flow in normal, open-chest dogs, at the site of application of potassium or serotonin to the surface of the epicardial coronary artery [57]. In 1983, we succeeded in developing coronary spasm by giving histmine or serotonin to atherosclerotic miniature swine [26,58]. The details of our animal model will be described in the next section. Other investigators gave vasoactive substances, such as thio-thromboxane A_2 or the potassium channel blocker, tetraethylammonium, into the coronary circulation of normal rabbits with the development of diffuse coronary spasm [59,60]. Other animal studies which demonstrated transient decreases in coronary flow, ST changes, or increase in coronary resistance do not prove the development of coronary spasm in epicardial coronary arteries.

Characteristics of Our Animal Model of Coronary Spasm

We designed an animal model which fulfills the definition of coronary spasm in the narrowest sense, and then attempted to clarify the role of spasm in various stages of ischemic heart disease.

We first attempted to develop an experimental animal model in which spasm could be produced in what appeared to be normal epicardial large coronary arteries, with or without mild atherosclerosis, and that the provoked coronary spasm would result in acute myocardial ischemia.

Studies in Dogs

We began to examine the effects of alpha adrenergic stimuli on change in the outer diameter of the epicardial large coronary arteries of normal dogs, since clinical studies have emphasized the role of alpha adrenergic receptor activation [61,62]. However, increased activity of the alpha-adrenergic receptor by phenylephrine and norepinephrine after beta-blocking and/or somatic nerve stimulation under beta-blockade caused only a 3% reduction in diameter of the canine epicardial coronary artery, despite an elevation in systemic pressure [63]. When we gave ergonovine to atherosclerotic dogs, we found a significant augmentation of coronary vasoconstriction of the atherosclerotic site [64], but this increased vasoconstriction was not sufficient to fulfill our definition of coronary spasm. We then started to work with swine, since induction of hypercholesterolemia and atherosclerosis is easier and faster in this species.

Our First Series of Studies of Coronary Spasm in Mildly Atherosclerotic Miniature Swine

In 1983 we provoked coronary spasm by giving histamine or serotonin intravenously or into the coronary arteries of atherosclerotic miniature swine. The findings were similar to those seen angiographically in patients with variant angina, as shown in Fig. 1 [26]. In our first study, the swine were given a diet containing 2% cholesterol for 3–6 months, following endothelial denudation using a balloon catheter to produce a localized atherosclerotic lesion. A reduction in the luminal diameter of a large epicardial coronary artery was measured by coronary arteriography before and after administration of various vasoactive agents. In our animal studies, coronary artery spasm was defined as transient excess vasoconstriction which subsides either spontaneously or following the administration of nitroglycerin, and is characterized by a greater than 75% transient reduction of the epicardial large coronary artery diameter, compared with the diameter 2 min after intravenous administration of 20 µg/kg nitroglycerin. Regional myocardial ischemia in the territory perfused by the spastic coronary artery was confirmed by observing ECG changes during balloon-induced obstruction of the same coronary artery.

Histamine study. Histamine, (200 µg) given into a coronary artery in the presence of cimetidine (60 mg/kg intravenously), an antagonist of the histamine H_2 receptor, caused an approximately 20% reduction in the luminal diameter before denudation, but an average reduction of 85% 3 months after denudation, in pigs on a high cholesterol diet (Fig. 1). The rate of induction of coronary

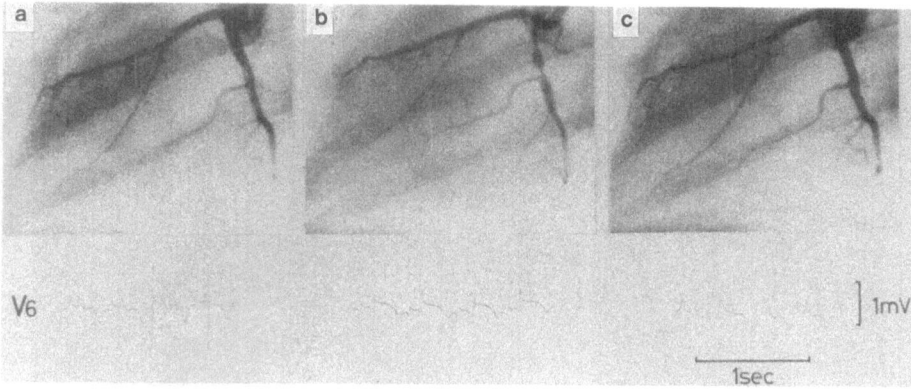

Fig. 1. Coronary arteriograms in miniature swine fed an atherogenic diet: **a** before and **b** after intracoronary administration of histamine (200 μg) in the presence of cimetidine administered intravenously (60 mg/kg), and **c** after intravenous administration of nitroglycerin (20 μg/kg). Coronary artery spasm, with associated electrocardiographic ST-segment elevation in the precordial lead, was induced at multiple sites on the circumflex branch of the left coronary artery. Skip-lesions of coronary atherosclerosis were observed histologically at the corresponding sites of spasm. (from [26] with permission of the American Association for the Advancement of Science)

spasm at the end of 3 months was approximately 80% (7 of 9 pigs). Ischemia-related ECG change was noted after histamine administration in 4 of 7 pigs. The histamine-induced spasm was prevented by diphenhydramine, a H_1-receptor blocker [26].

Serotonin study. Hyperconstriction of large coronary artery was also provoked by serotonin (30 μg/kg intravenously, or 60 μg into the coronary artery), but the induction rate and degree of reduction of intraluminal diameter were relatively low when compared with the findings seen with histamine [26]. Coronary spasm of the denuded site was never observed in this animal model after giving phenylephrine (20 μg into the coronary artery or 3 μg/kg intravenously) [26,58] or thiothromboxane A_2 (an analogue of thromboxane A_2). Indomethacin (2 mg/kg) or prostacyclin (50 ng/kg min^{-1}) failed to prevent the histamine-induced coronary spasm [65]. Thus, an imbalance between thromboxane A_2 and prostacyclin cannot explain the occurrence of coronary spasm.

Histologic study and serum cholesterol level. Histologic studies demonstrated that mild intimal thickening accompanied by proliferation of smooth muscle cells and accumulation of foam cells occurred in the spastic portion of the left coronary artery, as shown in [Fig. 2] [26]. However, platelet adhesion, thrombus, and hemorrhage were not present in the arterial wall, and endothelial cells covered the intimal surface. Topographic correlation was suggested between the site of spasm and the site of localized intimal thickening of the denuded portion [58]. Serum total cholesterol of these animals increased from control 64 mg/dl to 225 mg/dl, after 3 months. HDL-cholesterol had not significantly changed at the end of 3 months

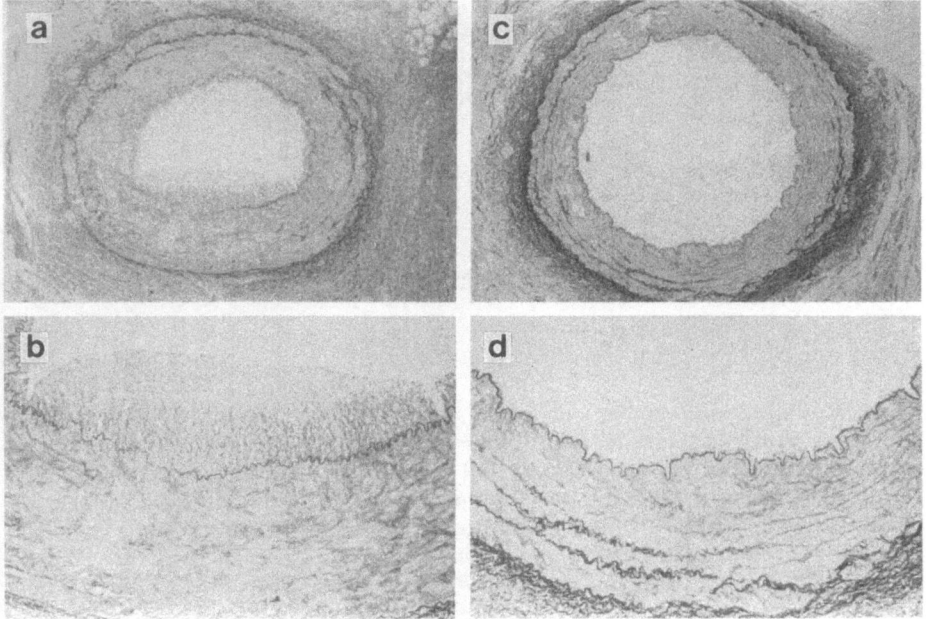

Fig. 2. Histology of the left circumflex coronary artery: **a, b** site of induced coronary artery spasm (×32) **c, d** anterior descending coronary artery (controls) (×80). Atherosclerotic changes characterized mainly by intimal thickening are distinct, yet these changes were not apparent angiographically. [From [26] with permission of the American Association for the Advancement of Science)

Close topological correlation between the site of coronary spasm and spontaneous or induced coronary atherosclerosis in swine. In 5 of 36 consecutively examined miniature swine, coronary spasm was provoked by simple intracoronary administration of histamine before endothelial denudation (Fig. 3). There was spontaneous development of atherosclerosis at the site of spasm in normocholesterolemic animals [66]. This would suggest that the presence of an atherosclerotic lesion is more important than hypercholesterolemia as a cause of spasm.

The close topological correlation between the site of spontaneous and/or induced atherosclerotic changes and the site of provoked spasm was noted, although the lesions were not angiographically visible. We noted that undenuded coronary arteries from hypercholesterolemic animals (average 222 mg/dl) fed a high cholesterol diet constricted more in response to histamine compared with the arteries of normocholesterolemic pigs [66].

Our Improved Models for Coronary Spasm in Moderately
Atherosclerotic Swine
Induction of coronary spasm by serotonin, ergonovine, and hyperventilation in a swine model. We then developed a more advanced atherosclerotic lesion by

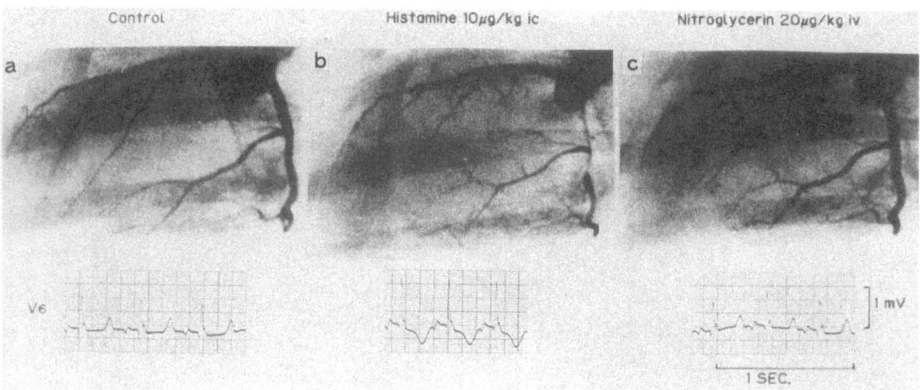

Fig. 3. Coronary angiograms and electrocardiograms obtained in **a** controls and **b** after administration of histamine and **c** after nitroglycerin, in a representative pig with spontaneous atherosclerosis. Tubular-type vasoconstriction occurred along the major trunk of the left circumflex coronary artery after intracoronary administration of 10 μg/kg histamine. The angiogram obtained after nitroglycerin administration shows no organic lesion of the left anterior descending or circumflex arteries. Electrocardiogram recorded at V_6 shows ST depression after histamine administration. [From [66] with permission of the American Heart Association)

feeding semi-synthetic diets containing high fat, 2% cholesterol, and 1.1% sodium cholate for several months, with an aim to elucidate the role of coronary spasm in acute myocardial infarction and/or progression of atherosclerotic narrowing. During the experiments, we were surprised to find no evidence of localized spasm, but diffuse hyperconstriction to histamine was noted not only at the denuded site but also at the undenuded site in hypercholesterolemic (335 mg/dl) miniature swine [67]. We then decided to apply local X-ray irradiation in addition to endothelial denudation and a high cholesterol diet in order to produce localized advanced atherosclerosis [27]. Angiography revealed a normal coronary arteriogram before giving serotonin, but after its administration (10 μg/kg) a transient typical spasm associated with ischemic ECG changes was induced in the coronary arteries at the X-ray irradiated site, as shown in [Fig. 4]. Only 30% luminal reduction was noted at the non-irradiated site [27]. The induction rate of coronary spasm increased to 90% or greater. However, the reduction in luminal diameter induced by histamine was approximately 60–70% and was rarely associated with acute myocardial ischemia [67]. We have been able to improve the animal model of coronary spasm by the following methods. A semi-synthetic diet containing high fat, 2% cholesterol, and 1.1% sodium cholate was given. Endothelial denudation was rigorously performed, and X-ray irradiation was performed 2 and 3 months after the start of the high cholesterol diet. Because miniature swine often develop ventricular fibrillation and die suddenly after severe coronary spasm, we titrated the minimum amount of serotonin (1 μg/kg or less intracoronary administration to cause approximate-

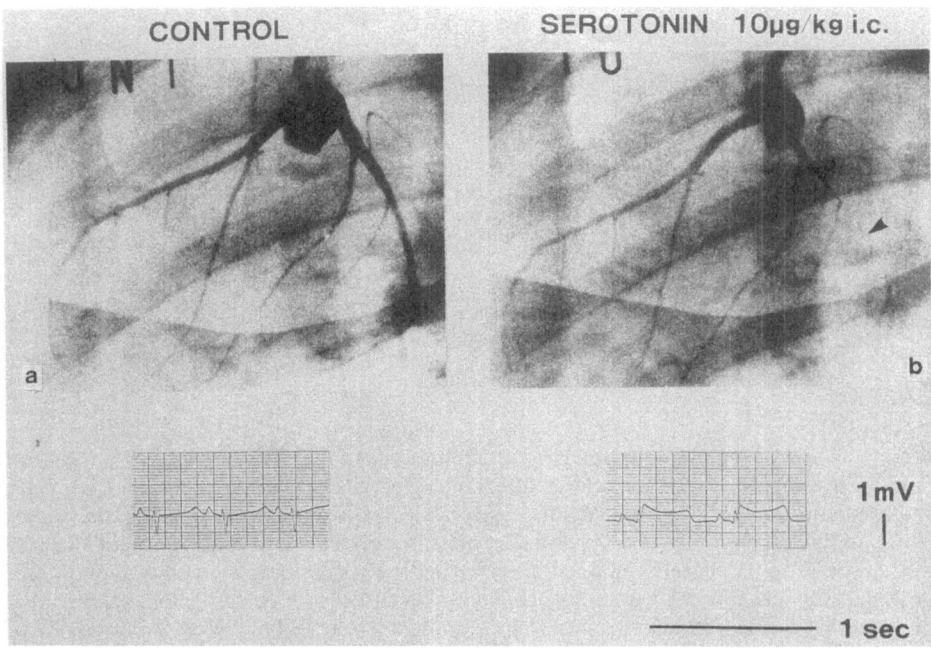

CONTROL   SEROTONIN 10µg/kg i.c.

a   b

1mV

1 sec

Fig. 4. Left coronary angiograms (left anterior oblique projection) of **a** control and **b** repetitive induction of coronary spasm with serotonin. Subtotal stenosis was noted along the left circumflex coronary artery. *Arrow* indicates site of severe spasm. During coronary spasm, ECG-ST elevation was recorded. (From [27] with permission of the American Heart Association)

ly 90% narrowing angiographically, without causing ventricular fibrillation, yet causing acute myocardial ischemia. In some pigs, even a single induction of coronary spasm caused subendothelial and intraplaque hemorrhage, and prolonged spasm caused complete occlusion and acute myocardial infarction [68]. We also noted that coronary spasm was provoked by serotonin, ergonovine, and hyperventilation only in the X-ray irradiated swine without endothelial denudation and/or a high cholesterol diet [69].

Induction of intramural hemorrhage, and progression of organic stenosis after coronary spasm: In patients with variant angina, acute myocardial infarction often occurred in the area supplied by spastic coronary artery. However, the relation between transient coronary spasm and persistent coronary occlusion has not been vigorously studied. To elucidate the relation between coronary spasm and acute myocardial infarction, an animal model with advanced coronary atherosclerosis and also inducible coronary spasm must be designed. In one of the previously mentioned improved animal models for spasm, 10 Goettingen-type miniature swine were fed a diet containing high fat, 2% cholesterol, and 1.1% sodium cholate for several months, and endothelial denudation and X-ray irradiation were performed twice [27]. In 4 pigs (Group A) transient coronary spasm was provoked only once by intracoronary injection of serotonin (10 µg/

kg. i.c.), and in 6 pigs (group B) serotonin was given 5 times at 5 min intervals. As shown in (Fig. 4), typical coronary spasm associated with ischemic ECG changes was induced in the area irradiated after serotonin administration, in all pigs of groups A and B. Forty minutes after the final provocation of spasm, the animals were killed and morphological studies performed. Light microscopy revealed intramural hemorrhage at the spastic site, in all pigs in the repetitive

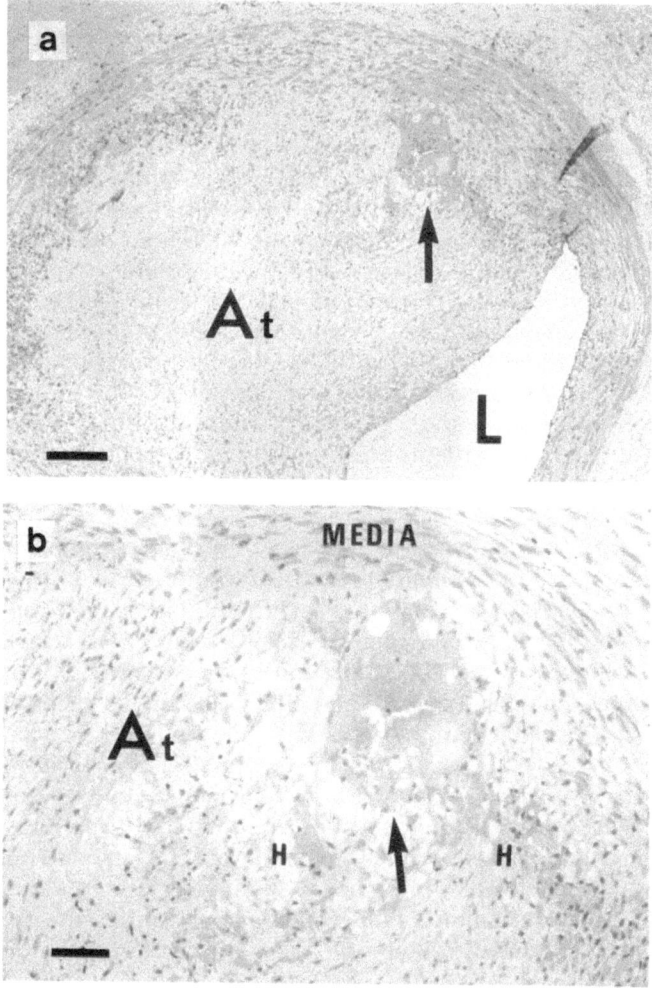

Fig. 5. Microscopic features of vessel with induced hemorrhage in group B. **a** Low-power magnification view. Bar represents 200 μm. *At* atheroma; *L* lumen. *Arrow* indicates the site of intramural hemorrhage, which is further magnified in B. **b** High-power magnification view. Bar represents 80 μm. Red blood cells (*arrow*) are preserved in their original shape, and there is no evidence of a histiocytic reaction or hemosiderin deposition. H, hyaline substance and red blood cells. From [27] with permission of the American Heart Association)

Table 1. Percent area stenosis along the spastic vessel (morphological determination)

	Proximal (%)	Spastic site (%)	Distal (%)
Group A	15 ± 7	23 ± 5	13 ± 6
		P < 0.01	
Group B	13 ± 4	56 ± 7	16 ± 3

Data are presented as Mean ± SEM. Samples of proximal and distal sites were taken 5 to 10 mm from the most severely spastic site. (From [27] with permission of the American Heart Association)

spasm group (group B), but none in group A (Fig. 5). Intramural hemorrhage was present inside the intima in all 6 pigs. However, rupture of the fibrous cap of the atherosclerotic plaque, fresh thrombus, and histological communication between the lumen and intramural hemorrhage were not observed.

Luminal organic stenosis, as assessed histologically at the most severely spastic site and at sites 5–10 mm proximal and distal to the spastic site was measured. The percentage area of stenosis at the intramural hemorrhagic site was significantly greater than that at the spastic site, seen with a single induction of spasm (group A), and was also greater than that at the proximal and distal sites in both groups A and B. [Table 1] Thus, coronary spasm can induce intramural hemorrhage, with progression to organic coronary stenosis. Histological findings suggested that the intramural hemorrhage observed in the present model was recent.

Scanning electron microscopy revealed that the non-spastic site in both groups A and B was covered with endothelial cells. In addition to the appearance of the intercellular bridges between endothelial cells at the spastic sites in both groups A and B, squeezing of endothelial cells, gaps between endothelial cells, and adhesion of white blood cells were present at the spastic site in group B (Fig. 6)[27]. These results suggest a cause-and-effect relation between coronary spasm and structural disarrangement, such as intramural hemorrhage and subsequent progression of coronary stenosis. The increased luminal narrowing at the hemorrhagic site may be potentially hazardous and result in a more disastrous chain of events, such as plaque rupture, luminal thrombosis, and acute myocardial infarction. These studies are now in progress at our laboratory. It has been shown that activated endothelial cells secrete a leukocyte adhesion inhibitor, such as interleukin-8 [70]. Further study of the interaction between leucocytes and endothelial cells in the large atherosclerotic coronary artery is needed.

Mechanisms of Coronary Spasm

In addition to our swine model, the mechanisms of coronary artery spasm in humans require much further study.

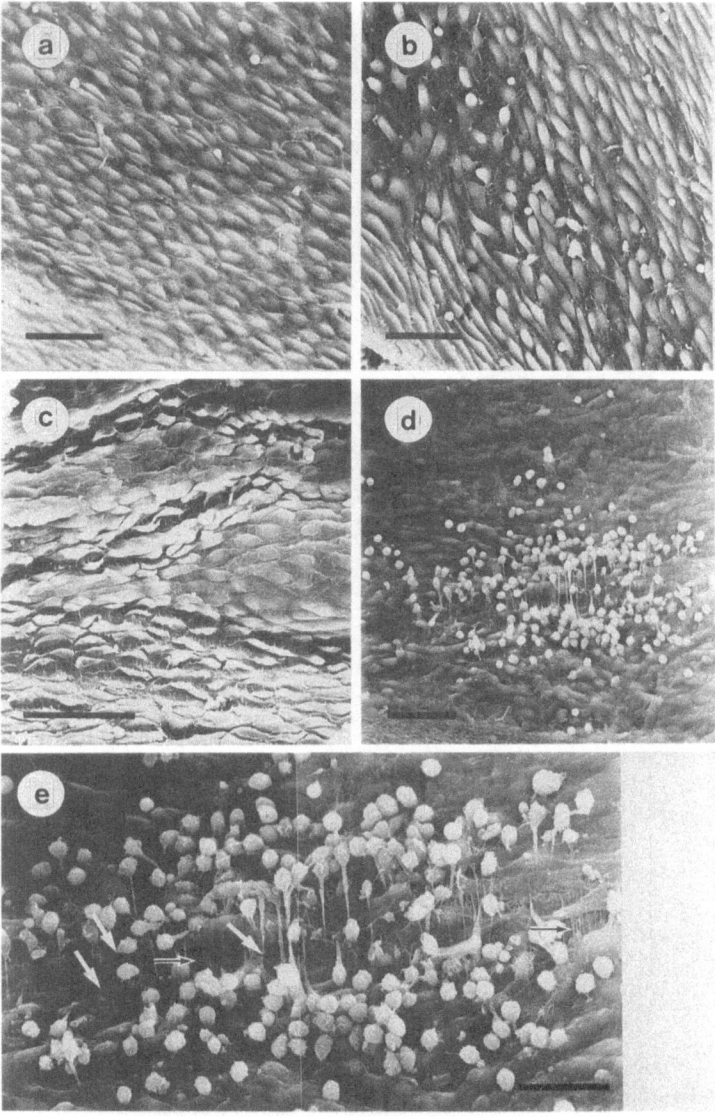

Fig. 6. Scanning electron micrographs of the luminal surfaces of coronary arteries. **a** non-spastic site; **b** spastic site after a single induction of spasm; **c** spastic site after five episodes of coronary spasm; **d** leukocyte adhesions at the spastic site after five repetitive inductions; **e** higher magnification view of *d*. Numerous fine bridges are seen between endothelial cells in *b*, *c*, and *d*. In *e*, *white arrows* indicate gaps between endothelial cells, and *black and white arrows* are endothelial bridges. *Bar* represents 50 μm in *a-d* and 25 μm in *e* [From [27] with permission of the American Heart Association)

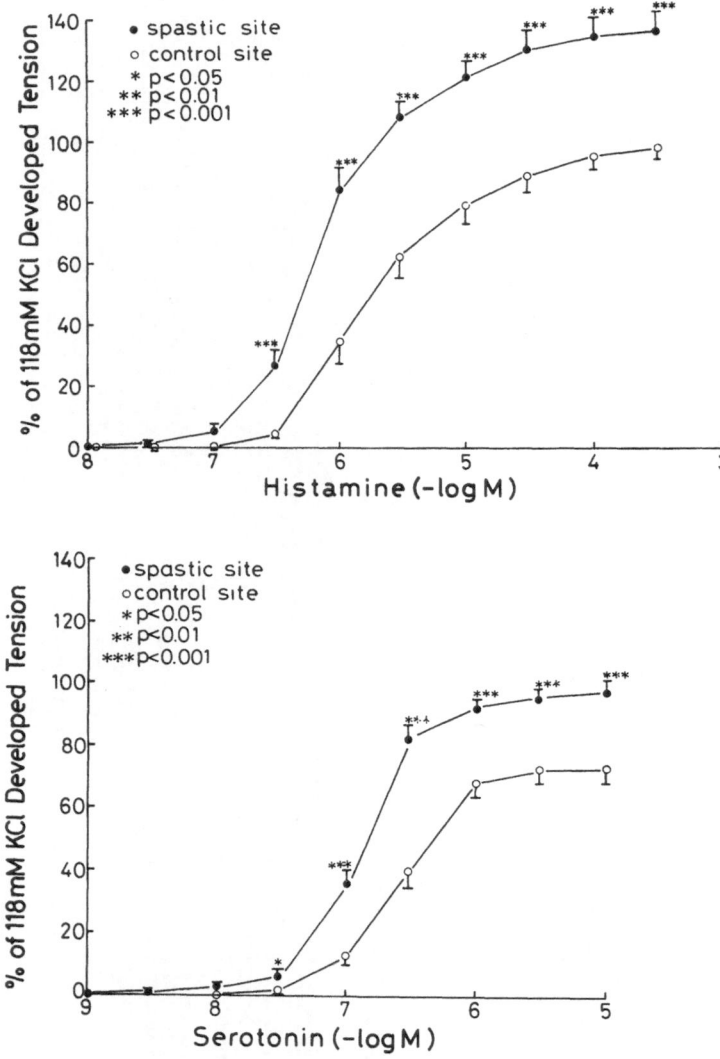

Fig. 7. Dose-response relation to histamine (*top*) and serotonin (*bottom*) in control and spastic vessels. Data are presented as mean + SEM for 14 and 13 preparations with histamine and serotonin, respectively. (From [75] with permission of the American Heart Association)

Coronary Atherosclerosis and Coronary Arterial Spasm

The atherosclerotic coronary artery should not be viewed as a passive rigid tube, but rather as a dynamic system modulated by neurohumoral stimuli and/or blood elements to which it may exhibit a variable sensitivity [71]. Coronary spasm almost always develops at the atherosclerotic site, although the extent of angiographic coronary stenosis varies widely. Spasm rarely occurs in humans younger than 20 years. The clinical findings regarding close topographical relation between the site of spasm and atherosclerosis show a good parallel to our swine model [26,27,66]. Augmented constrictor responses to serotonin and other vasoactive substances were noted in isolated atherosclerotic arteries of laboratory animals and of humans [72,73]. However, the presence of coronary atherosclerosis does not always mean that coronary spasm will occur. The type of atherosclerotic lesions which are prone to coronary spasm is unknown. In addition to coronary atherosclerosis, constrictor substances, such as acetylcholine, histamine and serotonin from nerve terminals, mast cells or blood platelets, and an amplifier such as increase of blood pH after hyperventilation, and a decrease in pressure would be required for spasm to occur. MacAlpine's geometric theory [52] was not supported by quantitative measurements in patients with variant angina [74], and our first series of studies on the swine model of coronary spasm [66]. Events that make a segment of coronary artery wall hyperreactive to constrictor stimuli must be considered among the putative mechanisms of coronary spasm. In our swine model, an increased contraction induced by histamine and serotonin was noted in isolated atherosclerotic coronary spastic segments, as shown in Fig. 7 [75]. The isolated atherosclerotic coronary segments at the spastic site showed a selective attenuation of endothelium-dependent vasodilation induced by serotonin, as shown in Fig. 8 [75]. However, the findings of endothelium-dependent relaxation of atherosclerotic arteries are controversial; such vasodilation is attenuated in atherosclerotic human coronary arteries [76], whereas the magnitude of endothelium-mediated vasodilation after histamine and Substance P was similar in a senior adult and in a 2-month-old male [77]. Endothelium-dependent relaxation is restored after regression of atherosclerosis, despite the presence of intimal thickening [78]. The mechanism of this impairment of endothelium-dependent relaxation in the atherosclerotic portion is no doubt related to a decrease in production of endothelium-dependent relaxing factor (EDRF), impairment of diffusion of EDRF, attenuated response of smooth muscle cells, and degradation of EDRF after release. We have recently found not only a reduction in EDRF release, but also increased degradation of EDRF and its reversal by superoxidase in the WATANABE heritable atherosclerotic aortic media [79]. We also found that an increased number of histaminergic receptors and/or augmentation of signal transduction, but not Ca-sensitivity of the contractile proteins in atherosclerotic medial muscle cells cause histamine-induced hypercontraction in our swine model [80]. However, the characteristics of the atherosclerotic lesion particularly susceptible to coronary spasm remain to be defined. In patients with vasospastic angina, local coronary hyperactivity was noted in response to ergonovine [81,82], acetylcholine [83], pilocarpine [84,85], methacholine [84,86], histamine [87], and to an increase in blood pH after hyperventilation [88]. In our atherosclerotic animal model, serotonin, ergonovine, histamine, and hyperventilation provoke coronary spasm similar to that seen in patients with variant angina.

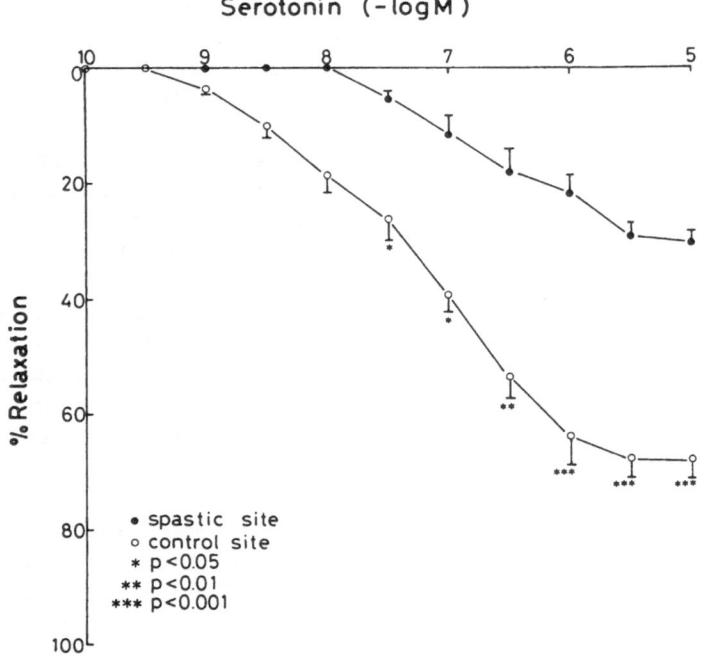

Fig. 8. Dose-response relation to serotonin for study of vascular relaxation in control and spastic portions precontracted by $PGF_{2\alpha}$. At doses greater than 3×10^{-7} M, serotonin produced less relaxation at the spastic site than at the control site. Data are mean + SEM for 5 preparations. (From [75] with permission of the American Heart Association)

Neurohumoral Factors

Neural Factors

Coronary spasm can occur in patients following total cardiac denervation produced by stripping of the great vessels or by autotransplantation [89–91]. However, cardiac sympathetic denervation, plexetomy, or total cardiac denervation reduced the rate of anginal attack in patients with variant angina although coronary spasm persisted [89,90,92]. The number of patients who have undergone cardiac denervation therapy for coronary spasm is too few to make any definite conclusion. In our animal model, coronary spasm provoked by histamine in situ was reproduced in vitro at the same site in the isolated perfused pig heart, as shown in Fig. 9 [93]. This finding may exclude the role of neural factors, at least in this swine model.

Adrenergic system. Traditionally, sympathetic nerve stimulation or norepinephrine infusion was believed to cause metabolically-mediated vasodilation of the coronary artery. This vasodilation competes with alpha-adrenergic vasoconstriction, and is minimized by pretreatment with beta-blockade [94–96]. Ricci et al. [62] reported that spontaneous coronary spasm was signaled by early,

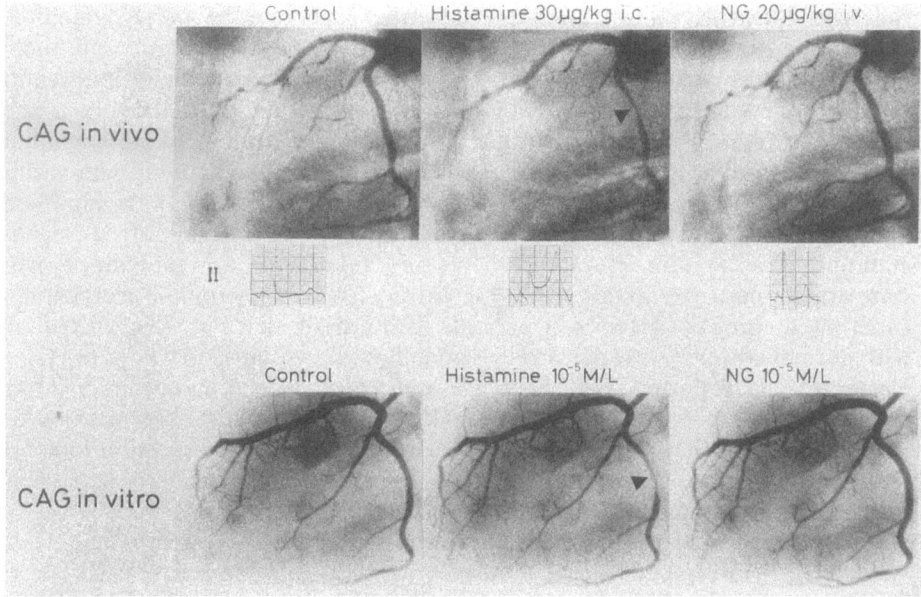

Fig. 9. Representative coronary angiograms of miniature pig, in vivo (*upper panel*) and in vitro (*lower panel*). *Arrow* indicates the site of coronary spasm. ECG lead II was recorded in vivo. The estimated concentration of histamine (10 µg/kg) in the intracoronary bolus was a maximum of 5.2×10^{-5} M. *CAG* coronary arteriography. (From [93] with permission of the American Heart Association)

transient prolongation of the QT interval, associated with increased cardiac sympathetic nerve activity, and that anginal attacks were relieved by phentolamine given intravenously or by oral phenoxybenzamine. Yasue et al. [61] reported adrenergic receptor-mediated coronary spasm in patients with variant angina, however, other studies discounted the importance of adrenergic stimulation in the pathophysiology of coronary spasm [97,98]. The lack of elevation of plasma catecholamine at the onset of ST segment elevation [99], and failure of prazosin and phentolamine to reduce anginal attacks in patients with variant angina [97,98] were also noted in a double-blind randomized trial.

Beta-blockade decreased large coronary artery diameter in laboratory animals [100], and sometimes accelerated anginal attack in patients with variant angina [61,99,101]. The possibility of localized alpha adrenergic-stimulation of epicardial coronary arteries by inflammation and other factors must be given attention [102]. If sympathetic nerve dysfunction persists, episodic neurogenic coronary spasm may occur at sites of platelet aggregation or increased serotonin [103].

Parasympathetic nervous system. The role of the parasympathetic nervous system in the regulation of coronary circulation remains controversial. The effects of cholinergic stimulation on the large coronary artery via neural activation of the parasympathetic nervous system are not clear, although it is known that the epicardial coronary artery is richly innervated with postganglionic cholinergic nerves, mainly in the adventitia [104]. No cholinergic endothelial innervation has been demonstrated [71], but cholinergic constrictor innervation involving a medial muscarinic receptor was demonstrated [105,106].

Several investigators demonstrated that the administration of methacholine, pilocarpine, and acetylcholine provoked coronary spasm in patients with variant angina [83–86]. The proposal that cholinergic stimulation of the sympathetic nervous system is responsible for anginal attack was not supported because cholinergic stimulation results in decreased release of norepinephrine from sympathetic nerve terminals [107,108]. Intracoronary injection of acetylcholine constricted coronary arteries of patients with atherosclerosis, while in patients with normal coronary arteries dosedependent mild vasodilation was noted [109]. Yasue et al. reported that acetylcholine constricted the coronary artery responsible for a variant anginal attack, but not the coronary arteries not responsible, and that pretreatment with atropine prevented induction of spasm [83]. These results suggest an important role for the parasympathetic nervous system in the pathogenesis of coronary spasm. However, this study does not necessarily prove that there is a neural effect by the parasympathetic nervous system, since atropine will act non-selectively on most muscarinic receptors. Studies on the direct effects of parasympathetic nerve activation on coronary tone were limited to the dog, and the results are controversial [105]. We do not know the mechanism of activation of cholinergic nerves in the coronary artery. It is unknown whether neurogenically liberated acetylcholine causes the release of EDRF in any significant amount. Furchgott and Zawadski demonstrated that damage or removal of the endothelial cells from isolated rabbit aorta reversed the relaxation induced by acetylcholine to vasoconstriction [110]. However, isolated normal human coronary artery obtained at autopsy and that of cattle, pigs, and sheep contracted in response to acetylcholine, independent of the presence or absence of an intact endothelium [105,111,112]. Human coronary arteries freshly obtained from recipient hearts during transplant showed a vasoconstrictive response to a muscarinic agonist [113]. Toda and Okamura obtained evidence that muscarinic receptors in the endothelium of human coronary arteries do not play an important role in the generation of relaxing factor [77].

Excitation of smooth muscle coronary artery-mediated muscarinic receptors may result from a G-protein-stimulated increase in the turnover of the phosphatidylinositol system, which increases mobilization of intracellular Ca, and directly increases the non-selective conductance of cations, causing a Ca-influx and leading to muscle contraction, as occurs in the gastric antrum [114]. Studies of receptor binding, a possible fluctuation of autonomic receptors, muscarinic receptor subtypes, and the levels of C-AMP and C-GMP in the coronary artery still need to be performed.

Humoral Factors (Autacoids and Platelets)

Histamine. Histamine infusion induced coronary spasm in patients with suspected variant angina [87], and also in atherosclerotic miniature swine [26].

In isolated human coronary artery, severely atherosclerotic coronary artery segments were supersensitive to histamine [113]. Histamine is synthesized in the arterial wall [115–117], and its concentration and the responsiveness of coronary arteries is increased in patients with cardiac disease [118]. In patients with peripheral vascular disease, the contents of histamine of leucocytes and platelets are elevated [119], and histamine promotes labilization of platelet granules by phorbol esters [120]. A remarkably increased number of adventitial mast cells containing histamine which is released in response to various stimuli was noted in coronary arteries of a patient with vasospastic angina, compared with the findings in patients with coronary artery disease and sudden death but no spasm, or in normal controls [121]. However, this observation does not rule out the occurrence of mast cell accumulation secondary to spasm, and the roles of mast cells in the pathogenesis of coronary spasm remain unknown [71]. Histamine is a potent vasoconstrictor, and causes contraction of human epicardial coronary arteries in vitro, through actions on the H_1 receptor [113,122]. In our swine model, histamine induced coronary spasm via the H_1-receptor, similar to that occurring in human coronary arteries [26]. In humans as well as in atherosclerotic miniature swine, the H_2 blocker, cimetidine. tends to augment H_1-receptor-mediated coronary spasm [26,123]. The potential of H_1-receptor antagonists to limit coronary spasm in patients with variant angina has not been substantiated.

Serotonin and platelets. Ergonovine has been widely used to induce coronary spasm in humans, and ergonovine-induced coronary spasm has been reported to closely resemble spontaneously occurring coronary arterial spasm [124]. Vasoconstriction produced by ergonovine was independent of neural control extrinsic to the heart, because the response to ergonovine was the same in heart transplant recipients lacking extrinsic neural control, and in patients with normally innervated hearts [53]. Ergonovine is classified as a serotonin agonist [125], and does not produce alpha-adrenergic receptor stimulation [126]. Henry and Yokoyama first reported that isolated aorta undergoing atherosclerotic changes was supersensitive to ergonovine, predominantly by a serotonergic mechanism [73]. Cohen demonstrated that aggregating platelets cause contraction of isolated canine coronary artery largely by the release of serotonin, and that the released serotonin accumulates in coronary adrenergic endings, and following release from nerves as a false transmitter, the amine can activate the serotonergic receptor on smooth muscle cells, and reverse the action of the adrenergic nerve from dilator to constrictor [103].

Aliquots of coronary sinus plasma from patients with coronary artery disease caused constriction of canine coronary artery rings, and only the serotonin antagonist, methiothepin prevented this constriction. However, plasma from patients with no coronary artery disease evoked only endothelium-dependent relaxation [127]. These results suggest a close relation between vasoconstrictor activity and the presence of coronary artery disease. Difference in the concentration of serotonin between coronary sinus and aorta was greater in patients with coronary heart disease compared with patients without significant coronary artery disease [128]. Earlier studies demonstrated reduced platelet survival [129,130], and increased platelet adhesiveness and aggregability in vitro in subjects with atherosclerosis [131,132]. Platelet-specific proteins, such as beta-thromboglobulin and platelet factor IV were increased in plasma of patients with coronary artery disease [133,134]. A correlation was also seen between increased circulating platelet aggregates and unstable angina or acute myocardial infarction, but not with stable angina or control patients with non-ischemic chest pain [135]. It is still not clear whether enhanced platelet reactivity is a cause of acute ischemic event, particularly in cases of vasospastic angina. Despite the evidence that platelet function seems important, inhibiting platelet aggregation with cycloxygenase inhibitors, such as aspirin [136,137], indomethacin [137], thromboxane A_2 synthetase inhibitors (OKY-046) [138], prosatcyclin [139], and the serotonin-S_2 antagonist (Ketanserin) [140,141] did not significantly decrease the frequency or severity of coronary spasm in patients with vasospastic angina. A specific S_1-antagonist is not available for clinical trials. Evidence from clinical and animal studies linking platelets or serotonin to coronary spasm is scanty and conflicting.

In Vivo and In Vitro and Species Differences

Most thorough studies on the coronary circulation have been performed in dogs. Species differences and local differences in the artery are known to exist [112,142]. The large epicardial coronary artery in the dog dilates in the presence of acetylcholine, both in vivo and in vitro, whereas the human coronary artery contracts to acetylcholine in vitro [77,105,111], but does not always contract in vivo in patients with coronary heart disease [109]. Isolated vessels often do not respond in the same way in vitro as in situ, even in the same species. Whether these differences reflect variations in neuroeffector organization and signal transduction systems or basic differences in the excitation-contraction coupling system of smooth muscle cells remains unknown.

Conclusion

We demonstrated in miniature pigs that coronary spasm associated with acute myocardial ischemia and sudden death can be induced in the presence of coronary atherosclerosis by the administration of serotonin, histamine, ergono-vine, and by hyperventilation, events similar to the spasm occurring in patients with variant angina. Coronary spasm could be reproduced in the isolated perfused heart. Intramural hemorrhage and progression of coronary stenosis, both considered to be causal lesions of acute myocardial infarction, could be produced by induction of coronary spasm. In isolated coronary artery segments from atherosclerotic swine, abnormal functions of endothelial cells and medial muscles at the spastic site were noted. Thus, the atherosclerotic swine model of coronary spasm is useful to elucidate the pathophysiology of ischemic heart disease. However, the definition of coronary spasm, as well its in vitro and in vivo mechanisms require further study.

Acknowledgements. The author thanks M. Ohara for helpful comments and I. Tomizuka for secretarial assistance. This work was supported by grants from the Ministry of Education, Science, and Culture and from the Ministry of Health and Welfare of Japan.

References

1. MacAlpine RN (1980) Coronary arterial spasm; A historical perspective. J Hist Med Allied Sci 35: 288–311
2. Osler W (1910) The Lumleian lectures on angina pectoris II. Lancet I: 839–844
3. Keefer CS, Resnik WH (1928) Angina pectoris: A syndrome caused by anoxemia of the myocardium. Arch Intern Med 41: 769–807
4. Blumgart HL, Schlesinger MJ, Davis D (1940) Studies on the relation of the clinical manifestations of angina pectoris, coronary thrombosis and myocardial infarction to the pathologic findings. Am Heart J 19: 1–91
5. Zoll PM, Wessler S, Blumgart HL (1951) Angina pectoris; a clinical and pathological correlation. Am J Med 11: 331–357
6. Proudfit WL, Shirey EK, Sones FM (1966) Selective cine coronary arteriography; Correlation with clinical findings in 1,000 patients. Circulation 33: 901–910
7. Prinzmetal M, Kennamer R, Merliss R, Wada T, Bor N (1959) Angina pectoris I: A variant form of angina pectoris preliminary report. Am J Med 27: 375–388
8. Guazzi M, Polese A, Fiorentini C, Magrini F, Bartorelli C (1971) Left ventricular performance and related hemodynamic changes in Prinzmetal's variant angina pectoris. Br Heart J 33: 84–94
9. Scherf D, Cohen J (1974) "Variant " angina pectoris. Circulation 49: 787–789
10. Epstein SE, Redwood DR, Goldstein RE, Beiser GD, Rosing DR, Glancy DL, Reis RL, Stinson EB (1971) Angina pectoris: pathophysiology, evaluation and treatment. Ann Intern Med 75: 263–296
11. Maseri A, Minno R, Chierchia S, Marchesi C, Pesola A, L'Abbate A (1975) Coronary artery spasm as a cause of acute myocardial ischemia in man. Chest 68: 625–632
12. Maseri A, Severi S, DeNes M, L'Abbate A, Chierchia S, Marzilli M, Ballestra AM, Parodi O, Biagini A, Distante A (1978) "Variant" angina: One aspect of continuous spectrum of vasospastic myocardial ischemia. Pathogenetic mechanisms, estimated incidence, clinical and coronarographic findings in 138 patients. Am J Cardiol 42: 1019–1035
13. Maseri A, Chierchia S (1982) Coronary artery spasm: Demonstration, definition, diagnosis, and consequences. Prog Cardiovasc Dis. 25: 169–192
14. Shimokawa H, Nagasawa K, Irie T, Egashira S, Egashira K, Sagara T, Kikuchi Y, Nakamura M (1988) Clinical characteristics and long term prognosis of patients with variant angina. A comparative study between western and Japanese populations. Int J Cardiol 18: 331–349
15. Nakamura M (1987) Task report: Basic and clinical studies of coronary spasm (in Japanese).
J Jap. Soc Intern Med 76: 1171–1187
16. Hiramori K (1985) Clinical problems related to vasospastic angina (in Japanese). J Jap. Coll Angiol 25: 399–403
17. Nakamura M, Takeshita A, Nose Y (1987) Clinical characteristics associated with

myocardial infarction, arrhythmias and sudden death in patients with vasospastic angina. Circulation 75: 1160–1116

18. Specchia G, de Servi S, Falcone C, Bramucci E, Angoli L, Mussini A, Marinoni GP, Montemartini C, Bobba P (1979) Coronary arterial spasm as a cause of exercise induced ST-segment elevation in patients with variant angina. Circulation 59: 948–954

19. Yasue H, Omote S, Takizawa A, Nagao M, Miwa K, Tanaka S (1979) Exertional angina pectoris caused by coronary arteria spasm; effects of various drugs. Am J Cardiol 43: 647–652

20. Bertrand ME, Lablanche JM, Tilmant PY (1980) Frequency of provocated coronary artery spasm in 273 patients with chest pain (abstract). Am J Cardiol 45: 390

21. Koiwaya Y, Torii S, Takeshita A, Nakagaki O, Nakamura M (1982) Postinfarction angina caused by coronary arterial spasm. Circulation 65: 275–280

22. Maseri A, L'Abbate A, Baroldi G, Chierchia S, Marzilli M, Ballestra AM, Severi S, Parodi O, Biagini A, Distante A, Pesola A (1978) Coronary vasospasm as a possible cause of myocardial infarction: A conclusion derived from the study of "preinfarction" angina. N Engl J Med 299: 1271–1277

23. Oliva PB, Breckinridge JC (1977) Arteriographic evidence of coronary arterial spasm in acute myocardial infarction. Circulation 56: 336–374

24. Marzilli M, Goldstein S, Trirella MG, Palumbo C, Maseri A (1980) Some clinical considerations regarding the relation of coronary vasospasm to coronary atherosclerosis; A hypothetical pathogenesis. Am J Cardiol 45: 882–886

25. Lown B, Desilva RA (1980) Is coronary arterial spasm a risk factor for coronary atherosclerosis? (editorials), Am J Cardiol 45: 901–903

26. Shimokawa H, Tomoike H, Nabeyama S, Yamamoto H, Araki H, Nakamura M, Ishii Y, Tanaka K (1983) Coronary artery spasm induced in atherosclerotic miniature swine. Science 221: 560–562

27. Nagasawa K, Tomoike H, Hayashi Y, Yamada A, Yamamoto T, Nakamura M (1989) Intramural hemorrhage and endothelial changes in atherosclerotic coronary artery after repetitive episodes of spasm in X-ray irradiated hypercholesterolemic pigs. Circ Res 65: 272–282

28. Dewood MA, Spores J, Notske R, Mouser LT, Burroughs R, Golden MS, Lang HT (1980) Prevalence of total coronary occlusion during the early hours of transmural myocardial infarction. N Engl J Med 303: 897–902

29. Benacerraf A, Scholl JM, Achard F, Tonnelier M, Lavergne G (1983) Coronary spasm and thrombosis associated myocardial infarction in a patient with nearly normal coronary arteries. Circulation 67: 1147–1150

30. Vincent GM, Anderson JL, Marshall HW (1983) Coronary spasm producing coronary thrombosis and myocardial infarction. N Engl J Med 309: 220–223

31. Conti RC, Feldman RL (1985) Acute myocardial infarction: Thoughts about pathogenesis and treatment. Mod Con Cardiovasc Dis 54: 35–38

32. Spann JF (1983) Changing concepts of pathophysiology, prognosis and therapy in acute myocardial infarction. Am J Med 74: 877–886

33. Rentrop P, Blanke H, Karsch KR, Kaiser H, Kostering H, Leitz K (1981) Selective intracoronary thrombolysis in acute myocardial infarction and unstable angina pectoris. Circulation 63: 307–317

34. Leinbach RC, Gold HK (1982) Coronary angiography during acute myocardial infarction: A search for spasm. Am Heart J 103: 768–772

35. Scotti TM, Hackel DB (1985) Heart. In: Kissane J (ed) Anderson's pathology, 8th edn. Mosby, St. Louis, pp 560–662

36. Hatanaka K, Ito T, Shiomi M, Yamamoto A, Watanabe Y (1987) Ischemic heart disease in the WHHL rabbit; A model for myocardial injury in genetically hyperlipidemic animals. Am Heart J 113: 280–288

37. Falk E (1983) Plaque rupture with severe pre-existing stenosis precipitating coronary thrombosis; Characteristics of coronary atherosclerotic plaques underlying fatal occlusive thrombi. Br Heart J 50: 127–134

38. Constantinides P (1984) Atherosclerosis-A general survey and synthesis. Surv Syn Path Res 3: 477–498

39. Wartman WB (1938) Occlusion of the coronary arteries by hemorrhage into their walls. Am Heart J 15: 459–470

40. Paterson JC (1938) Capillary rupture with intimal hemorrhage as a causative factor in coronary thrombosis. Arch Pathol 25: 474–487

41. Singh RN (1984) Progression of coronary atherosclerosis; Clues to pathogenesis from serial coronary arteriography. Br Heart J 52: 451–461

42. Davies MJ, Thomas AC (1985) Plaque fissuring-the cause of acute myocardial infarction, sudden ischemic death, and crescendo angina. Br Heart J 53: 363–373

43. Conti RC (1985) Variant angina and coronary artery spasm. In: Connor WE, Bristow ID (eds) coronary heart disease. Lippincott, Philadelphia, pp 251–267

44. Maseri A, L'Abbate A, Chierchia S, Parodi O, Severi S, Biagini A, Distante A, Marzilli M, Ballestra AM (1979) Significance of spasm in the pathogenesis of ischemic heart disease. Am J Cardiol 44: 788–792

45. Oliva PB, Ridge W (1984) Unstable rest angina with ST-segment depression. Ann Intern Med 100: 424–440

46. Brown BG, Dodge HT (1982) Unstable angina; Guidelines for therapy based on the last decade of clinical observations. Ann Intern Med 97: 921–923

47. Epstein SE, Palmeri ST (1984) Mechanisms contributing to precipitation of unstable angina and acute myocardial infarction; implications regarding therapy. Am J Cardiol 54: 1245–1252

48. Buxon A, Goldberg S (1980) Refractory ergonovine induced coronary vasospasm; importance of intracoronary nitroglycerlin. Am J Cardiol 46: 329–334

49. Crevey BJ, Owen SF, Pitt B (1981) Irreversible coronary occlusion related to administration of ergonovine. Circulation 64: 853–856

50. Bertrand ME, Lablanche JM, Tilmant PY, Thieuleux FA, Delforge MR, Carre AG, Asseman P, Berzin B, Libersa C, Laurrent JM (1982) Frequency of provoked coronary artery spasm in 1089 consecutive patients undergoing coronary arteriography. Circulation 65: 1229–1306

51. Lange R, Reid M, Tresch D, Keelan M, Bernhard V, Coolidge G (1972) Nonatheromatous ischemic heart disease following withdrawal from chronic industrial nitroglycerin exposure. Circulation 46: 666–678

52. MacAlpine RN (1980) Contribution of dynamic vascular wall thickening to luminal narrowing during coronary arterial constriction. Circulation 61: 296–301

53. Cipriano PR, Guthaner DF, Orlick AE, Ricci DR, Wexler L, Silverman JF (1979) The effects of ergonovine maleate on coronary arterial size. Circulation 59: 82–89

54. Tomoike H, Egashira K, Yamada A, Hayashi Y, Nakamura M (1987) Leukotriene C4-and D4-induced diffuse peripheral constriction of swine coronary artery accompanied by ST elevation on the electrocardiogram; angiographic analysis. Circulation 76: 480–487

55. Muramatsu K, Tomoike H, Ohara Y, Egashira S, Nakamura M (to be published) Effects of endothelin on epicardial coronary diameter, coronary blood flow, ECG-ST changes and wall motion. Heart and Vessels

56. Gensini GG, di Giorgi S, Murad-Netto S, Black A (1962) Arteriographic demonstration of coronary artery spasm and its release after the use of a vasodilator in a case of angina pectoris and in the experimental animal. Angiology 13: 550–553

57. Perez JE, Saffitz JE, Gutierrez FA, Henry PD (1983) Coronary artery spasm in intact dogs induced by potassium and serotonin. Circ Res 52: 423–431

58. Shimokawa H, Tomoike H, Nabeyama S, Yamamoto H, Ishii Y, Tanaka K, Nakamura M (1985) Coronary artery spasm induced in miniature swine; angiographic evidence and relation to coronary atherosclerosis. Am Heart J 110: 300–310

59. Yui Y, Sakaguchi K, Susawa T, Hattori R, Takatsu Y, Yui N, Kawai C (1987) Thromboxane A_2 analogue induced coronary artery vasoconstriction in the rabbit. Cardiovasc Res 21: 119–123

60. Iwaki M, Mizobuchi S, Nakaya Y, Kawano K, Niki T, Mori H (1987) Tetraethylammonium induced coronary spasm in isolated perfused rabbit heart: A hypothesis for the mechanism of coronary spasm. Cardiov Res. 21: 130–139

61. Yasue H, Touyama M, Kato H, Tanaka S, Akiyama F (1976) Prinzmetal's variant form of angina as a manifestation of alpha-adrenergic receptor-mediated coronary artery spasm: documentation by coronary arteriography. Am Heart J 91: 148–155

62. Ricci DR, Orlick AE, Cipriano PR, Guthaner DF, Harrision DC (1979) Altered adrenergic activity in coronary arterial spasm; insight into mechanism based on study of coronary hemodynamics and the electrocardiogram. Am J Cardiol 43: 1073–1079

63. Nakamura M, Tomoike H, Ootsubo H, Sakai H, Noguchi K, Takeshita A, Kikuchi Y (1981) Constriction of the epicardial coronary artery induced by alpha-adrenergic stimulation. Basic Res Cardiol 76: 498–502

64. Kawachi Y, Tomoike H, Maruoka Y, Kikuchi Y, Araki H, Ishii Y, Tanaka K, Nakamura M (1984) Selective hypercontraction caused by ergonovine in the canine coronary artery under conditions of induced atherosclerosis. Circulation 69: 441–450

65. Shimokawa H, Tomoike H, Nabeyama S, Yamamoto H, Nakamura M (1985) Histamine-induced spasm not significantly modulated by prostanoids in a swine model of coronary artery spasm. J Am Coll Cardiol 6: 321–327

66. Egashira K, Tomoike H, Yamamoto Y, Yamada A, Hayashi Y, Nakamura M (1986) Histamine-induced coronary spasm in regions of intimal thickening in miniature pigs: roles of serum cholesterol and spontaneous or induced intimal thickening. Circulation 74: 826–837

67. Tomoike H, Hayashi Y, Egashira K, Yamada A, Nakamura M (1989) Effects of atherogenic diets on serum content of cholesterol, histamine-induced coronary constriction and morphological changes of the coronary artery (abstract). Jpn Circ J 53: 900

68. Kuga T, Tagawa A, Tomoike H, Nakamura M (1990) Abrupt onset of coronary spasm causes progress of organic stenosis, and prolonged coronary spasm induces acute myocardial infarction (abstract No.0814; in Japanese). Jpn Circ J (Suppl) 54: 206

69. Ohara Y, Tagawa H, Kuga T, Tomoike H, Nakamura M (1989) Augumentation of serotonin-induced coronary constriction in respiratory or metabolic alkalosis in miniature pigs (abstract). Circulation 80 (Suppl II): 234

70. Gimbrone Jr, Obin MS, Brock AF, Luis EA, Hass PE, Hebert CA, Yip YK, Leung DW, Kohr WJ, Darbonne WC, Bechtol KB, Baker JB (1989) Endothelial interleukin-8: A novel inhibitor of leukocyte-endothelial interactions. Science 246: 1601–1603

71. Bassenge E, Busse R (1988) Endothelial modulation of coronary tone. Prog Cardiovasc Dis 30: 349–380
72. Ginsburg R, Bristow MB, Davis K, Dibiase A, Billingham ME (1984) Quantitative pharmacologic responses of normal and atherosclerotic isolated human epicardial coronary arteries. Circulation 69: 430–440
73. Henry PD, Yokoyama M (1980) Supersensitivity of atherosclerotic rabbit aorta to ergonovine. J Clin Invest 66: 306–313
74. Freedman B, Richmond DR, Kelley DT (1982) Pathophysiology of coronary artery spasm. Circulation 66: 705–709
75. Yamamoto Y, Tomoike H, Egashira K, Nakamura M (1987) Attenuation of endothelium-related relaxation and enhanced responsiveness of vascular smooth muscle to histamine in spastic coronary arterial segments from miniature pigs. Circ Res 61: 772–778
76. Förstermann U, Mügge A, Alheid U, Haverich A, Frolich J (1988) Selective attenuation of endothelium-mediated vasodilation in atherosclerotic human coronary arteries. Circ Res 62: 185–190
77. Toda N, Okamura T (1989) Endothelium-dependent and-independent responses to vasoactive substances of isolated human coronary arteries. Am J Physiol 257: H988–995
78. Harrison DG, Armstrong ML, Freiman PC, Heisted DD (1987) Restoration of endothelium dependent relaxation by dietary treatment of atherosclerosis. J Clin Invest 80: 1808–1811
79. Tagawa H, Tomoike H, Nakamura M: Putative mechanisms of endothelium-depend relaxation of aorta in the prescence of atheromatous plaque in hereditary hyperlipidemic rabbits. Circ. Res., in press
80. Satoh S, Tomoike H, Mitsuoka W, Egashira S, Tagawa H, Kuga T, Nakamura M (1990) Smooth muscles from the spastic coronary artery segments shows hyper-contractility to histamine. Am J Physiol, 259: H9–H13
81. Stein I (1949) Observations on the action of ergonovine on the coronary circulation and its use in the diagnosis of coronary artery insufficiency. Am Heart J 37: 36–45
82. Schroeder JS, Bolen JL, Quint RA, Clark DA, Hayden WG, Higgins CB, Wexler L (1977) Provocation of coronary spasm with ergonovine maleate. New test with results in 57 patients undergoing coronary arteriography. Am J Cardiol 40: 487–491
83. Yasue H, Horio Y, Nakamura N, Fujii H, Imoto N, Sonoda R, Kugiyama K, Obata K, Morikami Y, Kimura T (1986) Induction of coronary artery spasm by acetylcholine in patients with variant angina: Possible role of the parasympathetic nervous system in the pathogenesis of coronary artery spasm. Circulation 74: 955–963
84. Yamamoto M, Katayama S (1968) Angina pectoris induced by marked vagotonic state (abstract). Jpn Circ J 32: 1856
85. Torii S, Araki Y, Sagara T, Kakimaru S, Kanaya H, Naito S, Nakagaki O (1974) Provocation of coronary spasm by subcutaneous administration of pilocarpine (in Japanese). Heart 6: 338–346
86. Endo M, Hirosawa K, Kaneko N, Hase K, Inoue Y, Konno S (1976) Prinzmetal's variant angina: Coronary arteriogram and left ventriculogram during anginal attack induced by methacholine. N Engl J Med 294: 252–255
87. Ginsburg R, Bristow MR, Kantrowitz N, Baim DS, Harrison DC (1981) Histamine provocation of clinical coronary artery spasm: Implications concerning pathogenesis of variant angina pectoris. Am Heart J 102: 819–822

88. Yasue H, Nagano M, Omoto S, Takizawa A, Miwa K, Tanaka S (1978) Coronary arterial spasm and Prinzmetal's variant form of angina induced by hyperventilation and tris-buffer infusion. Circulation 58: 56–62

89. Clark DA, Quint RA, Mitchell RL, Angell WW (1977) Coronary artery spasm; medical management, surgical denervation, and autotransplantation. J Thorac Cardiovasc Surg 73: 332–339

90. Bertrand ME, Lablanche JM, Tilmant PY, Ducloux G, Warembourg Jr H, Soots G (1981) Complete denevation of the heart (autotransplantation) for treatment of severe, refractory coronary spasm. Am J Cardiol 47(6): 1375–1378

91. Buda AJ, Fawles RE, Schroeder JS, Hunt SA, Cipriano PR, Stinson EB, Harrison DC (1981) Coronary artery spasm in the denervated transplanted human heart. Am J Med 70: 1144–1149

92. Betrin A, Pomar JL, Bourassa MG, Grondin CM (1983) Influence of partial sympathetic denevation on the results of myocardial revascularization in variant angina. Am J Cardiol 51: 661–667

93. Yamamoto Y, Tomoike H, Egashira K, Kobayashi T, Kawasaki T, Nakamura M (1987) Pathogenesis of coronary artery spasm in miniature swine with regional intimal thickening after balloon denudation. Circ Res 60: 113–121

94. Pitt B, Elliot EC, Gregg DE (1967) Adrenergic receptor activity in the coronary arteries of the unanesthetized dog. Circ Res 21: 75–84

95. Vatner SF, Higgins CB, Braunwald E (1974) Effects of norepinephrine on coronary circulation and left ventricular dynamics in the conscious dog. Circ Res 34: 812–823

96. Mohrman DF, Feigl EO (1978) Competition between sympathetic vasoconstriction and metabolic vasodilation in the canine coronary circulation. Circ Res 42: 79–86

97. Winniford MD, Filipchuk N, Hillis LD (1983) Alpha-adrenergic blockade for variant angina: A long term double-blind randomized trial. Circulation 67: 1185–1188

98. Chierchia S, Davies G, Berkenboom G, Crea F, Crean R, Maseri A (1984) Alpha-Adrenergic receptors and coronary spasm; An elusive link. Circulation 69: 8–14

99. Robertson D, Robertson RM, Niew AS, Oates JA, Friesinger GC (1979) Variant angina pectoris; Investigation of indexes of sympathetic nervous system function. Am J Cardiol 43: 1080–1085

100. Vatner SF, Hinze T (1983) Mechanism of constriction of large arteries by beta-adrenergic receptor blockade.Circ Res 53: 389–400

101. Nakamura M, Koiwaya Y (1979) Beneficial effect of diltiazem, a new antianginal drug, on angina pectoris at rest. Jpn Heart J 20: 613–621

102. Young MA, Knight DR, Vatner SF (1987) Autonomic control of large coronary arteries and resistance vessels. Prog Cardiovasc Dis 30: 211–234

103. Cohen RA (1985) Platelet-induced neurogenic coronary constriction due to accumulation of the false neurotransmitter, 5-hydroxytryptamine. J Clin Invest 75: 286–292

104. Denn MG, Stone HL (1976) Autonomic innervation of dog coronary arteries. J Appl Physiol 41: 30–35

105. Kalsner S (1989) Cholinergic constriction in the general circulation and its role in coronary artery spasm. Circ Res 65: 237–257

106. Hirsch EF, Borghard-Erdle AM (1961) The innervation of human heart. I. coronary arteries and the myocardium. Arch Pathol 71: 384–407

107. Kalsner S (1985) Coronary artery reactivity in human vessels; some questions and some answers. Fed Proc 44: 321–325

108. Vanhoutte PM, Verbeuren TJ, Webb RC (1981) Local modulation of adrenergic neuroeffector interaction in the blood vessel wall. Physiol Rev 61: 151–247

109. Ludmer PL, Selwyn AP, Shook TL, Wayne RR, Mudge GH, Alexander RW, Ganz P (1986) Paradoxical vasoconstriction induced by acetylcholine in atherosclerotic coronary arteries. N Engl J Med 315: 1046–1051

110. Furchgott RF, Zawadski JV (1980) The obligatory role of endothelial cells in the relaxation of arterial smooth muscle by acetylcholine. Nature 288: 373–376

111. Toda N (1983) Isolated human coronary arteries in response to vasoconstrictor substances. Am J Physiol 245: H937–H941

112. Kalsner S (1985) Cholinergic mechanisms in human coronary artery preparations: Implications of species differences. J Physiol 358: 509–526

113. Ginsburg R, Bristow MR, Harrison DC, Stinson EB (1980) Studies with isolated human coronary arteries. Chest 78 (Suppl): 180–186

114. Moummi G, Magous R, Strosberg D, Bali JP (1988) Muscarinic receptors in isolated smooth muscle cells from gastric antrum. Biochem Pharmacol 37: 1363–1369

115. EL-Ackad TM, Brody MJ (1975) Evidence for non-mast cell histamine in the vascular wall. Blood Vessels 12: 181–191

116. Garland CJ, Keatinge WR (1982) Constrictor actions of acetylcholine, 5-hydroxytryptamine and histamine on bovine coronary artery inner and outer muscle. J Physiol 327: 363–376

117. Howland RD, Spector S (1972) Disposition of histamine in mammalian blood vessels. J Pharmacol Exp Therap 182: 239–245

118. Kalsner S, Richards R (1984) Coronary arteries of cardiac patients are hyperreactive and contain stores of amines: A mechanism for coronary spasm. Science 223: 1435–1437

119. Gill DS, Barradas MA, Tonesca VA, Gracey L, Dandona P (1988) Increased histamine content in leukocytes and platelets of patients with peripheral vascular disease. Am J Clin Path 89: 622–626

120. McNicol A, Saxena SP, Brandes LJ, Gerrad JM (1989) A role of intracellular histamine in ultrastructural changes induced in platelets by phorbol esters. Arteriosclerosis 9: 684–689

121. Forman MB, Oates JA, Robertson D, Rabertson RM, Roberts, LJ II, Virmani R (1985) Increased adventitial mast cells in a patient with coronary spasm. N Engl J Med 313: 1138–1141

122. Ginsburg R, Bristow MR, Stinson EB, Harrison DC (1980) Histamine receptors in the human heart. Life Sci 26: 2245–2249

123. Shimokawa H, Okamatsu S, Taira Y, Nakamura M (1987) Cimetidine induces coronary artery spasm in patients with vasospastic angina. Can J Cardiol 3: 177–182

124. Curry RC, Pepine Jr CJ, Sabom MB, Conti CR (1979) Similarities of ergonovine-induced and spontaneous attacks of variant angina. Circulation 59: 307–312

125. Brazenor RM, Angus JA (1981) Ergometrine contracts isolated canine coronary arteries by serotonergic mechanism; no role for alpha adrenoceptors. J Pharmacol Exp Therap 218: 530–536

126. Holtz J, Held W, Sommer O, Kuhne G, Bassenge E (1982) Ergonovine induced constriction of epicardial coronary arteries in conscious dogs: Alpha-adrenoreceptors are not involved. Basic Res Cardiol 77: 278–291

127. Runbanyi GM, Frye RL, Holmes DR, Vanhoutte PM (1987) Vasoconstrictor activity of coronary sinus plasma from patients with coronary artery disease. J Am Coll Cardiol 9: 1243–1249

128. van den Berg EK, Schmitz JM, Benedict CR, Malloy CR, Willerson JT, Dehmer GJ (1989) Transcardiac serotonin concentration is increased in selected patients with limiting angina and complex coronary lesion morphology. Circulation 79: 116–124

129. Steele PP, Weily HS, Davise H, Genton E (1973) Platelet function studies in coronary artery disease. Circulation 48: 1194–1200

130. Ritchie JL, Harker LA (1977) Platelet and fibrinogen survival in coronary atherosclerosis; response of medical and surgical therapy. Am J Cardiol 39: 595–598

131. Horlick L (1961) Platelet adhesiveness in normal persons and subjects with atherosclerosis: Effect of high fat meals and anticoagulants on the adhesive index. Am J Cardiol 8: 459–470

132. Gormsen J, Nielsen JD, Andersen LA (1977) ADP-induced platelet aggregation in vitro in patients with ischemic heart disease and peripheral thromboatherosclerosis. Acta Med Scand 201: 509–513

133. Smitherman TC, Milam M, Woo J, Willerson JT, Frenkel UP (1981) Elevated beta-thromboglobulin in peripheral venous blood of patients with acute myocardial ischemia; direct evidence for enhanced platelet reactivity in vivo. Am J Cardiol 48: 395–402

134. Sobel M, Salzman EW, Davies GC, Handin RI, Sweeney J, Ploretz J, Kurland G (1981) Circulating platelet products in unstable angina pectoris. Circulation 63: 300–306

135. Schwartz MB, Hawiger J, Timmons S, Friesinger GC (1980) Platelet aggregates in blood from patients with ischemic heart disease. Thromb Haemost 43: 185–188

136. Chierchia S, DeCaterina R, Crea F, Patrono C, Maseri A (1982) Failure of thromboxane A_2 blockade to prevent attacks of vasospastic angina. Circulation 66: 702–705

137. Robertson RM, Robertson D, Roberts LJ, Maas RL, Fitzgerald GA, Friesinger GC, Oates JA (1981) Thromboxane A_2 in vasotonic angina pectoris: Evidence from direct measurements and in hibitor trials. N Engl J Med 304: 998–1003

138. Yui Y, Hattori R, Takatsu Y, Kawai C (1986) Selective thromboxane A_2 synthetase inhibition in vasospastic angina pectoris. J Am Coll Cardiol 7: 25–29

139. Chierchia S, Patrono C, Crea F, Ciabattoni G, Decaterina R, Cinotti GA, Distante A, Maseri A (1982) Effects of intravenous prostacyclin in variant angina. Circulation 65: 470–477

140. Freedman SB, Chierchia S, Rodriguez-Plaza L, Bugiardini R, Smith G, Maseri A (1984) Ergonovine induced myocardial ischemia: No role for serotonergic receptors. Circulation 70: 178–183

141. Caterin RD, Capeggiani C, L'Abbate A (1984) A double-blind, placebo-controlled study of ketanserin in patients with Prinzmetal's angina: Evidence against a role of serotonin in the genesis of coronary vasospasm. Circulation 69: 889–894

142. Vanhoutte PM, Shimokawa H (1989) Endothelium-derived relaxing factor and coronary spasm. Circulation 80: 1–9

CHAPTER 6

Endothelium-Derived Vasoactive Factors, Platelets and Coronary Disease

P.M. VANHOUTTE[1]

Summary. Endothelial cells can release both relaxing and contracting substances. The former include prostacyclin, endothelium-derived relaxing factor (EDRF, which most likely is nitric oxide, or a nitrosoderivative releasing nitric oxide, derived from 1-arginine), and endothelium-derived hyperpolarizing factor (EDHF, which possibly is a labile metabolite of arachidonic acid formed through the P-450 pathway). Possible endothelium-derived contracting factors (EDCF) include superoxide anions, thromboxane A_2, and the peptide endothelin. Endothelium-derived relaxing factor causes relaxation of vascular smooth muscle by activation of the soluble form of guanylate cyclase, which leads to an accumulation of cyclic GMP; it also reduces platelet adhesion and aggregation. The latter effect is synergistic with the inhibition evoked by prostacyclin. The release of endothelium-derived relaxing factor and prostacyclin plays a key role in the protective role of the endothelium against vasospasm and unwanted coagulation of blood. Indeed, thrombin and aggregating platelets are potent stimuli for the release of endothelium-derived relaxing factor. The platelet products responsible are the adenine nucleotides, ADP and ATP, which activate P_2-purinegic receptors on the endothelial cells, and 5-hydroxytryptamine (serotonin) which stimulates $5HT_1$-like serotonergic receptors. The response to serotonin, but not to the adenine nucleotides, is mediated by a pertussis toxin-sensitive mechanism. When endothelial cells regenerate, or are cultured, they selectively lose the pertussis toxin-sensitive mechanism of release, which results in a marked decrease in sensitivity to exogenous and platelet-released serotonin. As a consequence, the endothelial cells exhibit a considerably reduced response to aggregating platelets. This phenomemon, which can be exacerbated by hypercholesterolemia, favors ongoing platelet aggregation and vasospasm, and constitutes a first step toward atheroscerosis.

Introduction

It is now established beyond doubt that endothelial cells play a key role in the local control of vascular function. Indeed, since the original description of the

---

[1] Department of Medicine, Center for Experimental Therapeutics, Baylor College of Medicine, Houston, TX 77030–3498, USA

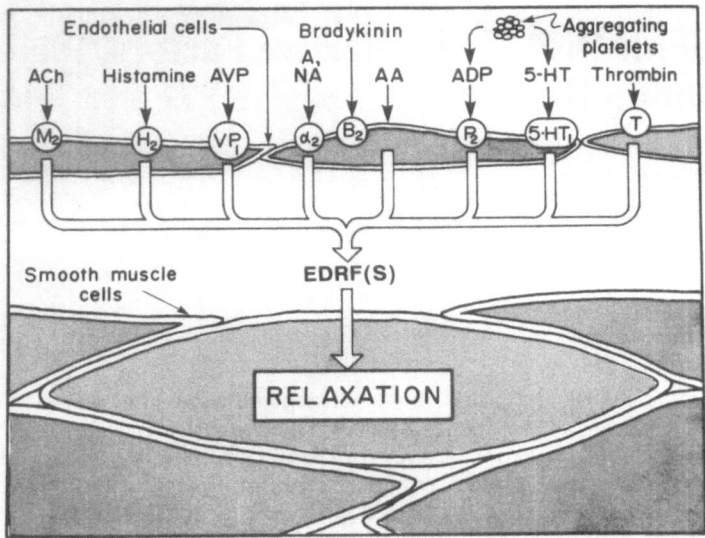

Fig. 1. Neurohumoral mediators which cause the release of endothelium-derived relaxing factor (*EDRF*) through activation of specific endothelial receptors (*circles*). EDRF can also be released independently of receptor-operated mechanisms by the calcium ionophore A23187 (not shown). A, adrenaline (epinephrine); AA, arachidonic acid; ACh, acetylcholine; ADP, adenosine diphosphate; α, alpha-adrenergic receptor; AVP, arginine vasopressin; B, kininreceptor; H, histaminergic receptor; 5-HT, serotonin (5-hydroxytryptamine); 5HT₁, serotonergic receptor; M, muscarinic receptor; NA, noradrenaline (norepinephrine); P, purinergic receptor; PGI₂, prostacyclin; T, thrombin receptor; VP, vasopressinergic receptor. (Reprinted with permission from CRC Press, Inc. Boca Raton, FL [7])

obligatory role played by the endothelium in the relaxations of isolated arteries evoked by acetylcholine [1], numerous investigators have demonstrated the occurrence of endothelium-dependent relaxation (Fig. 1) and contractions (Fig. 2) in a variety of isolated arteries, microvessels, and veins with a number of vasoactive substances. It becomes obvious that the endothelial cells also help to control the tone of the underlying smooth muscle in vivo [2–7]. This chapter focuses on the interaction between platelets and endothelial cells, in terms of the release of endothelium-derived vasoactive factors, how these interactions could contribute to physiological regulations, as well as to pathological responses of the blood vessel wall, and updates discussions of this topic [8–10].

Endothelium-Dependent Responses to Platelets

The Response

If quiescent isolated coronary arteries are exposed to aggregating platelets or to platelet-products, they contract vigorously. The contraction is reduced consider-

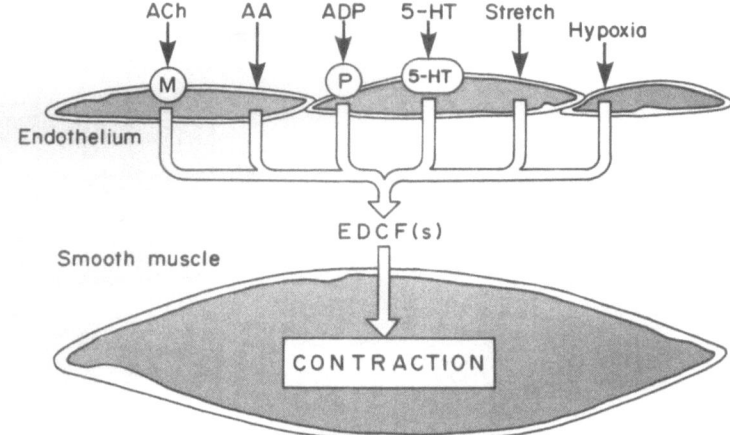

Fig. 2. A number of physiochemical stimuli, neurohumoral mediators, and the calcium ionophore A23187 can evoke endothelium-dependent contractions in certain blood vessels, presumably because they evoke the release of endothelium-derived contracting factor(s) (*EDCF*). AA, arachidonic acid; ACh, acetylcholine; ADP, adenosine diphosphate; 5-HT, serotonin (5-hydroxytryptamine), ➤ serotonergic receptor; M, muscarinic receptor; P, purinergic receptor. (Reprinted with permission from CRC Press, Inc. Boca Raton, FL [7])

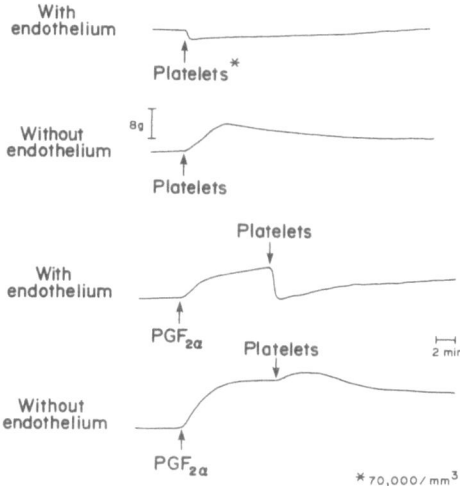

Fig. 3. Effect of the authors' platelets on the responsiveness of isolated canine arteries with and without endothelium, both under basal conditions (*upper two traces*) and during contractions to prostaglandin $F_{2\alpha}$ ($PGF_{2\alpha}$; *lower two traces*). When the platelet suspensions are added to the organ chambers, aggregation occurs immediately. This results in relaxation in the presence of endothelium, but causes contraction in its absence. These experiments demonstrate that aggregating platelets are a strong stimulus for the release of EDRF, and that the presence of the endothelium considerably inhibits the vasoconstrictor response to the platelet products. (From [13] with permission)

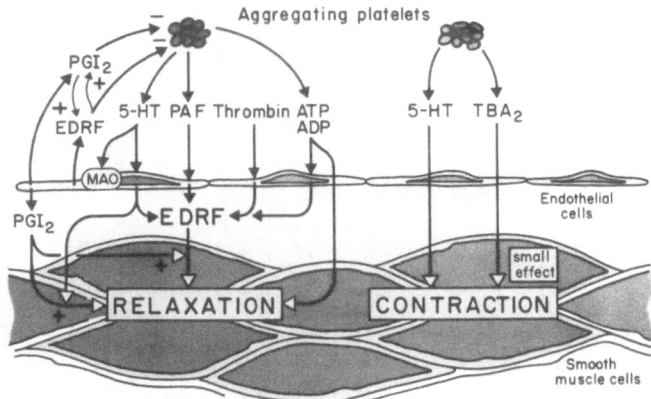

Fig. 4. Interactions between platelet products, thrombin and the endothelium. If the endothelium is intact, several of the substances released from the platelets [in particular, the adenine nucleotides (*ADP* and *ATP*) and serotonin (*5-HT*) and to a much lesser extent platelet activating factor (*PAF*)] cause the release of endothelium-derived relaxing factor (*EDRF*) and prostacyclin (*PGI*$_2$). The same is true for any thrombin formed. The released EDRF will relax the underlying vascular smooth muscle, opening up the blood vessel, and thus flushing the microaggregate away. It also will be released towards the lumen of the blood vessel to inhibit platelet adhesion of the endothelium, and synergistically with prostacyclin, inhibit platelet aggregation. In addition, monoamine oxidase (*MAO*) and other enzymes will break down the vasoconstrictor serotonin, limiting the amount of the monoamine that can diffuse toward the smooth muscle. Finally, (*right*) the endothelium acts as a physical barrier that prevents access of the vasoconstrictor platelet products serotonin and thromboxane A$_2$ (*TBA*$_2$) to the smooth muscle. These different endothelial functions play a key role in preventing unwanted coagulation of blood and vasospastic episodes in blood vessels with normal intima. If the endothelial cells are removed (e.g., by trauma) the protective role of the endothelium is lost locally, platelets can adhere and aggregate, and vasoconstriction follows; this contributes to the vascular phase of hemostasis. + , activation; − , inhibition. (From [10] with permission)

ably if the blood vessels contain functional endothelial cells (Fig. 3). If the smooth muscle of the arteries is first activated, prior to the exposure to the aggregating platelets, the latter cause further contractions in the absence of endothelium, but pronounced relaxations in the presence of endothelium (Fig. 3). The blunting effect of endothelium on platelet-induced contractions has been demonstrated in coronary and other arteries of several species, including humans [11–15]. It is likely that it contributes to the protective role of the endothelium in trying to avoid unwanted platelet-aggregation, coagulation of blood, and local vaso-constriction in blood vessels with an intact intima (Fig. 4) [8–10]. However, if the endothelial cells are removed, following trauma, the absence of inhibitory signals from the endothelium allows the full expression of the vasoconstrictor effect of the platelet-products (in particular serotonin and thromboxane A2) that underlie the vascular phase of hemostasis (Fig. 4).

The Mediators

Of the different products released by aggregating platelets, 5-hydroxytryptamine (serotonin) and the adenine nucleotides [adenosine diphosphate (ADP) and triphosphate (ATP)] play the major role in the platelet response. These compounds can evoke powerful endothelium-dependent relaxations which are mediated by 5-HT1-like receptors (Fig. 5) [16] and P2y-purinergic receptors (Fig. 6) [17] respectively. Platelet activating factor (PAF), which can cause endothelium-dependent relaxations at very high concentrations, contributes little to the platelet-induced response [9]. The contribution of adenine nucleotides and serotonin to platelet-induced, endothelium-dependent relaxations varies among species and between blood vessels. Thus, in the canine coronary artery, adenine nucleotides play a primary role [12,13], while in the porcine coronaries, serotonin predominates [14].

Serotonin and the adenine nucleotides do not utilize the same biochemical pathway to evoke endothelium-dependent relaxations (Fig. 7). This conclusion is derived from the observation that pertussis toxin, an inhibitor of certain G-proteins (in particular Gi-proteins), considerably blunts the endothelium-dependent relaxations evoked by serotonin (Fig. 8), but not those to ADP (Fig. 9) [18,19]. Data obtained from the coronary artery of pig shows that the pertussis toxin-sensitive component of the response to serotonin plays a crucial

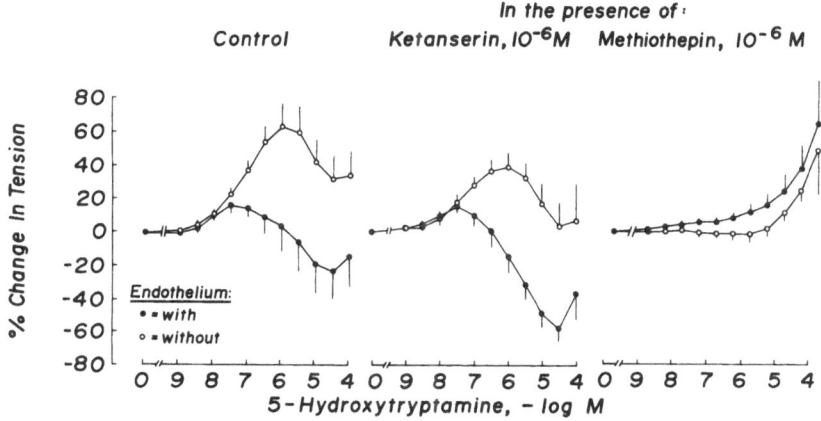

Fig. 5. In the canine coronary artery [contracted with prostaglandin F_{2a} (PF_{2a})] serotonin causes contractions in the absence of endothelium, and relaxations in its presence (*control, left*). Incubation with the $5HT_2$-serotonergic antagonist, ketanserin, modestly reduces the contraction in rings without endothelium, and augments the relaxation in rings with endothelium (*center*). The nonselective serotonergic antagonist methiothepin, shifts the contraction-response curve in rings without endothelium to the right and abolishes the difference between preparations with and without endothelium (*right*). These experiments demonstrate that the receptor mediating endothelium-dependent relaxations belongs to the $5HT_1$-family of serotonergic receptors. (From [12] with permission)

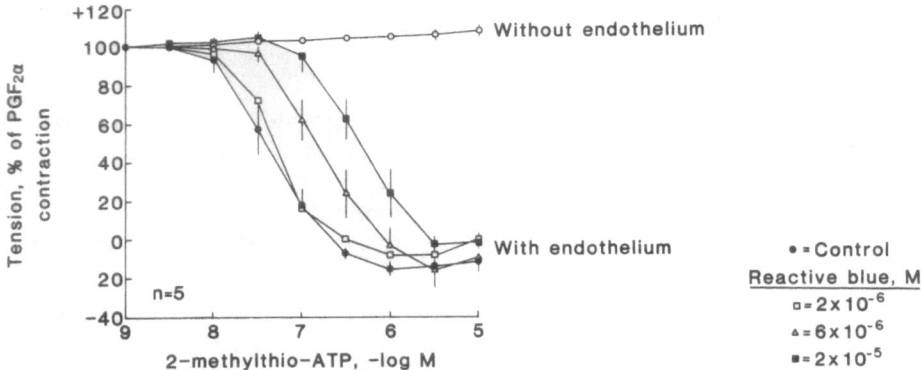

Fig. 6. The ATP-analogue 2-methylthio-ATP, which is a selective ligand for P_{2y}-purinoceptors, causes relaxations of the canine coronary artery which are strictly endothelium-dependent. The relaxations are antagonized in a competitive manner by the inhibitor of P_{2y}-purinoceptors, reactive blue. These experiments demonstrate that the purinoceptor which mediates the endothelium-dependent relaxation to the adenine nucleotides (ADP and ATP) belongs to the P_{2y} subtype. (From [17] with permission)

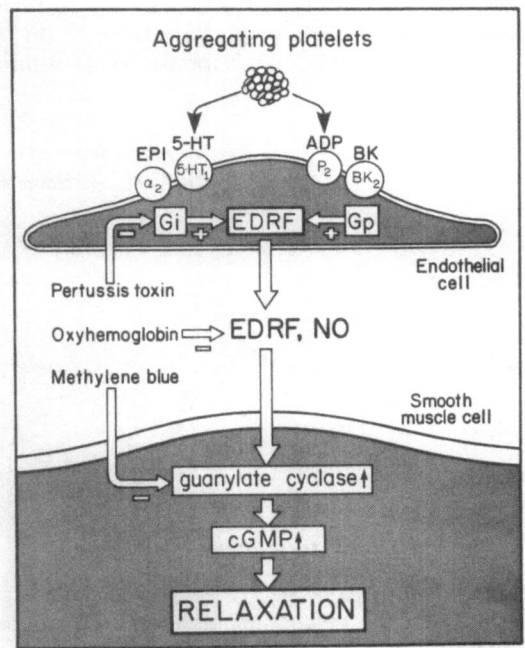

Fig. 7. The release of EDRF by endothelial cells involves at least two types of signal-transducing G-proteins, one of which is sensitive to pertussis toxin. Agonists at the endothelial cell membrane can use either or both pathways. α_2, α_2-adrenoceptor; ADP, adenosine diphosphate; BK, bradykinin; BK_2, kinin receptor; cGMP, cyclic GMP; EDRF, endothelium-derived relaxing factor; Gi and Gp, G_i and G_p-proteins; 5-HT, serotonin; 5-HT$_1$, serotonin receptor; NO, nitric oxide; P_2, P_{2y}-purinoceptor. (From [35] with permission)

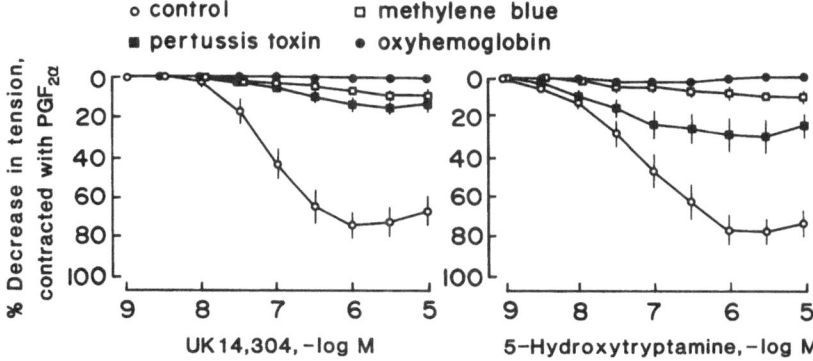

Fig. 8. In the porcine coronary artery the selective α_2-adrenergic agonist UK 14,304 (*left*) and serotonin (5-hydroxytryptamine, *right*) cause relaxations; although not shown, in both cases these relaxations are strictly endothelium-dependent, and are mediated by EDRF since they are prevented by oxyhemoglobin (an inactivator of the factor) and methylene blue (an inhibitor of soluble guanylate cyclase). Pertussis toxin, an inhibitor of certain G-proteins (in particular Gi-proteins) abolishes the response to UK 14,304, and markedly blunts that to serotonin. These experiments demonstrate that the endothelium-dependent response to both agonists involves pertussis toxin-sensitive G-proteins. (From [18] with permission)

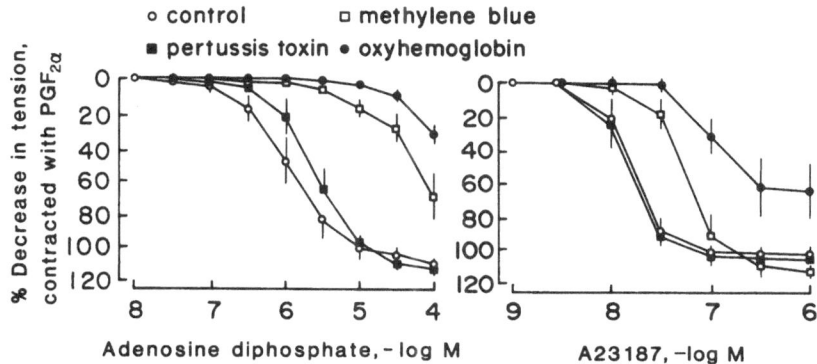

Fig. 9. Influence of the inhibitors of G-proteins; pertussis toxin, methylene blue, and oxyhemoglobin, on the endothelium-dependent relaxations produced by adenosine diphosphate (*left*) and the calcium ionophore A 23187 (*right*) in porcine coronary arteries. The arterial rings were contracted with prostaglandin $F_{2\alpha}$, and responses are expressed as percentage relaxation of the contraction. (From [18] with permission)

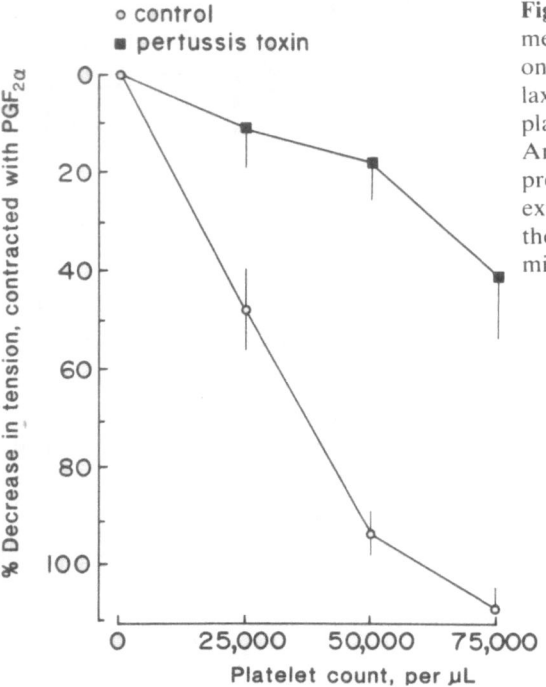

Fig. 10. Influence of pertussis toxin, methylene blue, and oxyhemoglobin on the endothelium-dependent relaxations produced by aggregating platelets in porcine coronary arteries. Arterial rings were contracted with prostaglandin $F_{2\alpha}$, and responses are expressed as percentage relaxation of the contraction. (From [18] with permission)

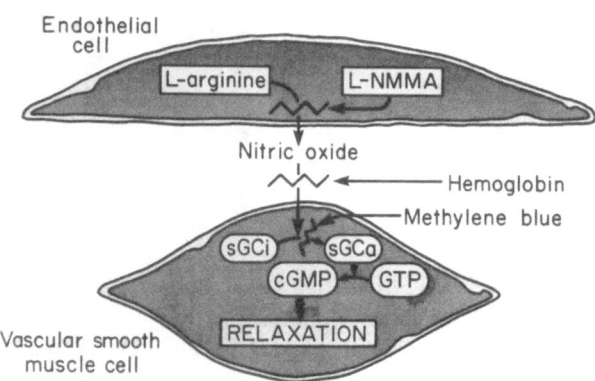

Fig. 11. The production of endothelium-derived nitric oxide from 1-arginine, and the inhibitory effect of L-N^G-monomethyl arginine (*L-NMMA*). The nitric oxide derived from L-arginine activates the soluble guanylate cyclase (*sGC*) in the smooth muscle, leading to the accumulation of cyclic guanosine monophosphate (*cGMP*). Hemoglobin inactivates nitric oxide, while methylene blue prevents the activation of soluble guanylate cyclase. i, inactive; a, active; GTP, guanosine triphosphate

role in the endothelium-dependent relaxations to the aggregating platelets (Fig. 10), possibly by amplifying the activation of the endothelial purinergic receptors by the platelet-derived adenine nucleotides.

The Endothelial Signal

Endothelial cells induce relaxation of the underlying smooth muscle by generating vasoactive substances. Besides prostacyclin [20], the most important endothelium-derived relaxing factor (EDRF) appears to be endothelium-derived nitric oxide (NO) [21–23] which is derived from the metabolism of L-arginine (Fig. 11) [24–26,6]. In addition, endothelial cells generate a hyperpolarizing and relaxing factor, the action of which can be prevented by ouabain (Fig. 12) [27,28]. Experiments performed with cultured endothelial cells suggest that ADP preferentially releases the ouabain-sensitive factor [29]. Experiments in intact, isolated porcine coronary arteries suggest that nitric oxide is the main mediator of the endothelium-dependent response to serotonin, ADP, and aggregating platelets. Indeed, scavengers of EDRF such as hemoglobin, and inhibitors of soluble guanylate cyclase such as methylene blue, profoundly inhibit these responses [19] (Figs. 8 & 9).

Endothelium-Derived Factors and Platelets

A major finding is the observation that endothelium-derived relaxing factor not only inhibits the underlying vascular smooth muscle, but also profoundly affects

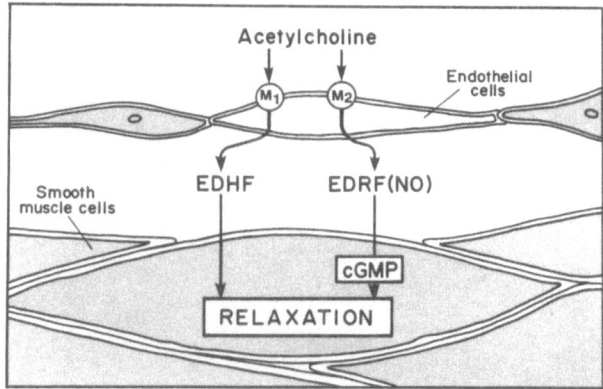

Fig. 12. Endothelial cells, when exposed to acetylcholine release two vasoactive factors. EDHF hyperpolarizes the cell membrane, thus initiating the relaxation and/or making the vascular smooth muscle more sensitive to the action of EDRF, which presumably is NO. The latter sustains the relaxation by entering the cell and activating soluble guanylate cyclase, which leads to an accumulation of cyclic GMP (*cGMP*). The muscarinic receptors (*M*) on the endothelial cell membrane which trigger the release of the two factors do not belong to the same subtype. (Reprinted with permission from CRC Press, Inc. Boca Raton, FL [7])

platelet-life. Thus, EDRF and nitric oxide inhibit the adhesion of platelets to the endothelium [30,31]; in addition EDRF inhibits platelet-aggregation, in particular if it is released together with prostacyclin [32,31]. Thus, at the interface between the blood and the blood vessel wall, the secretion of endothelium-derived products exerts a formidable negative feedback on the adhesion and aggregation of platelets (Fig. 4).

Pathology

A characteristic of blood vessels taken from animal models of vascular disease is to exhibit reduced endothelium-dependent relaxations. This phenomenon occurs very early. Thus, a few weeks after balloon-denudation of the endothelium in the porcine coronary artery, the endothelial cell lining is restored. Yet, the presence of these endothelial cells cannot significantly inhibit the vasoconstrictor response to aggregating platelets (Fig. 13) [14]. The defect in the response to the platelets is long-lasting, and can be attributed to a loss of the pertussis-toxin sensitive (presumably mediated through Gi-proteins) response to the platelet-derived serotonin (Figs. 14,15) [18,19]. This defect results in the

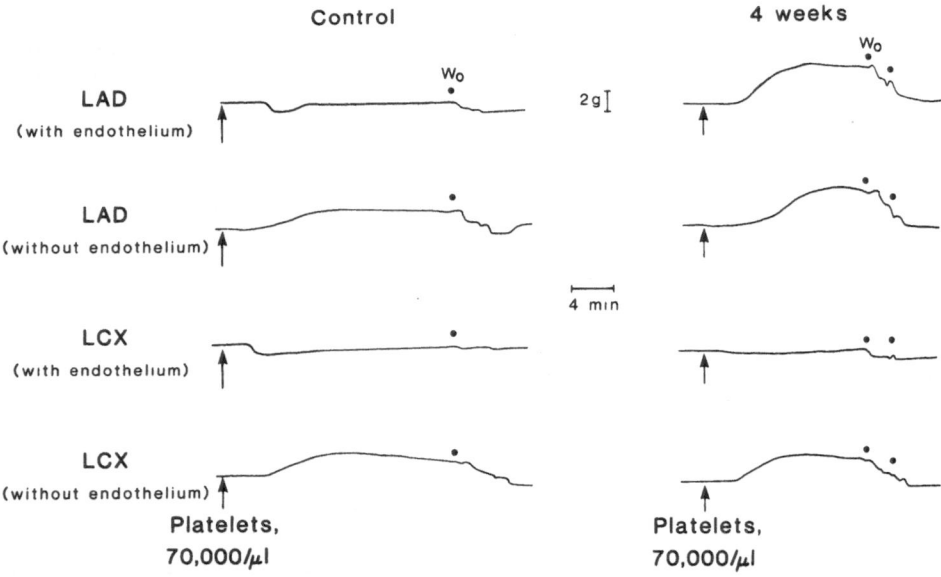

Fig. 13. Responses to aggregating platelets (*arrows*) in rings of porcine left anterior descending (*LAD*) and left circumflex (*LCX*) coronary arteries. *Left*: the blood vessels were taken from a control pig; in both cases, platelets evoke contractions only in rings without endothelium. *Right*: the rings were taken from a portion of the left anterior descending coronary artery that underwent balloon denudation 4 weeks prior to the experiment; the endothelium, unlike that of the circumflex coronary artery of the same animal, cannot prevent the contraction evoked by the platelet products. (From [14] with permission)

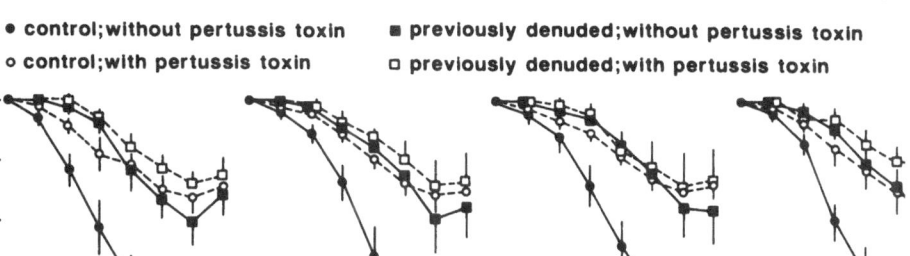

Fig. 14. Endothelium-dependent relaxations to serotonin in control porcine coronar arteries and in arteries that underwent balloon denudation 4, 8, 16, or 24 weeks prior t the experiment. The effect of pertussis toxin (an inhibitor of G-proteins) is also shown The degree of blunting of the response to serotonin observed in the previously denude arteries is comparable to that obtained acutely in normal arteries with pertussis toxin. I the previously denuded arteries, the toxin causes no further inhibition of relaxation to th monoamine. These observations strongly suggest that the regenerated endothelium ha lost the pertussis toxin-sensitive mechanism of release of endothelium-dependent relaxin; factor. In addition, this study illustrates that the abnormal response of the regenerate endothelium persists for at least 6 months. (From [36] with permission)

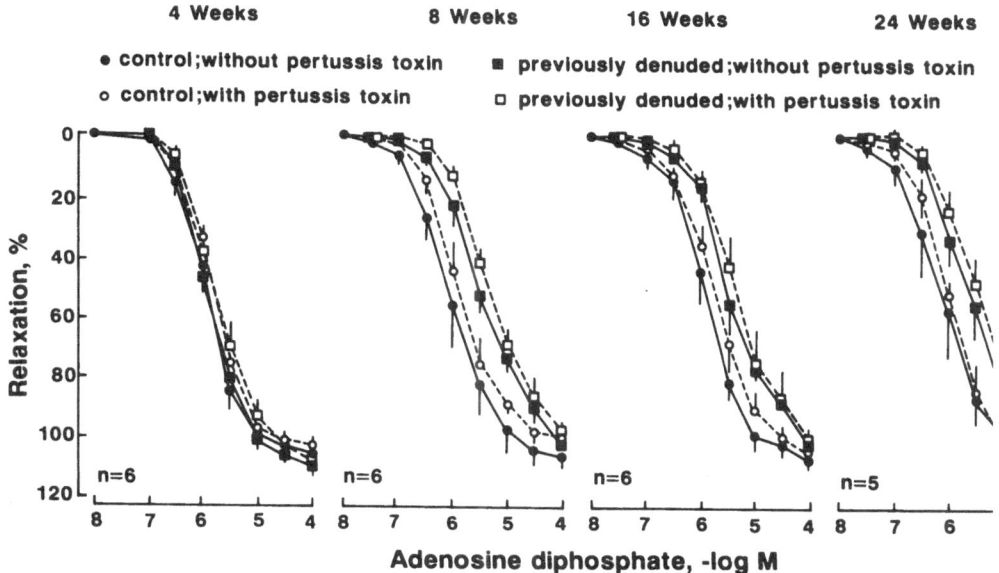

Fig. 15. Unlike the response to serotonin (Fig. 14) previous denudation of th endothelium and incubation with pertussis toxin does not, or only minimally, affect th endothelium-dependent relaxation to bradykinin. (From [36] with permission)

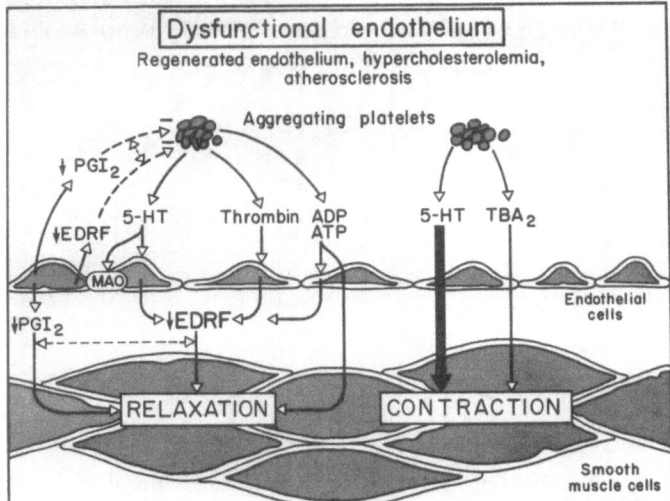

Fig. 16. Endothelium-dependent responses under pathologic conditions. The endothelium is dysfunctional in the regenerated state, hypercholesterolemia, and atherosclerosis, and releases less endothelium-derived relaxing factor (*EDRF*), however, the ability of the smooth muscle to contract is unaltered. As a result, the contractions predominate. In atherosclerosis, the production of both endothelium-dependent relaxing factor and prostacyclin (*PGI$_2$*) is reduced, and their synergistic actions against aggregating platelets may not occur. ADP, adenosine diphosphate; ATP, adenosine triphosphate; 5-HT, serotonin; MAO, monoamine oxidase; TBA$_2$, thromboxane A$_2$, inhibition; + , synergism. (From [10] with permission)

inappropriate reaction to aggregating platelets, favoring the occurrence of ongoing platelet-aggregation with the resulting release of not only the vasoconstrictors serotonin and thromboxane A2, but also platelet-derived growth factor(s) (Fig. 16) [10]. Thus, the localized absence of the vasodilator, anti-platelet, endothelium-dependent responses to aggregating platelets and possibly thrombin generated during the activation of the coagulation-cascade [33,34] could explain the repeated occurrence of vasospasm (and thrombosis) at sites of previous endothelial injury, and possibly explain the growth-process that leads to the intimal thickening characteristic of atherosclerosis.

References

1. Furchgott RF, Zawadzki JV (1980) The obligatory role of endothelial cells in the relaxation of arterial smooth muscle by acetylcholine. Nature 299: 373–376
2. Furchgott RF (1983) Role of the endothelium in responses of vascular smooth muscle. Circ Res 53: 557–573
3. Furchgott RF (1984) Role of endothelium in the responses of vascular smooth muscle

to drugs. Ann Rev Pharmacol Toxicol 24: 175–197

4. Vanhoutte PM, Rubanyi GM, Miller VM, Houston DS (1986) Modulation of vascular smooth muscle contraction by the endothelium. Annu Rev Physiol 48: 307–320

5. Bassenge E, Busse R (1987) Endothelial modulation of coronary tone. Prog Cardiovasc Dis 30: 349–380

6. Furchgott RF, Vanhoutte PM (1989) Endothelium-derived relaxing and contracting factors. FASEB J 3: 2007–2018

7. Lüscher TF, Vanhoutte PM (1990) The endothelium: Modulator of cardiovascular function. CRC, Boca Raton pp 1–228

8. Vanhoutte PM, Cohen RA (1983) The elusory role of serotonin in vascular function and disease. Biochem Pharmacol 32: 3671–3674

9. Vanhoutte PM, Houston DS (1985) Platelets, endothelium and vasospasm. Circulation 72: 728–734

10. Vanhoutte PM, Shimokawa H (1989) Endothelium-derived relaxing factor(s) and coronary vasospasm. Circulation 80: 1–9

11. Cohen RA, Shepherd JT, Vanhoutte PM (1983) Inhibitory role of the endothelium in the response of isolated coronary arteries to platelets. Science 221: 273–274

12. Houston DS, Shepherd JT, Vanhoutte PM (1985) Adenine nucleotides, serotonin and endothelium-dependennt relaxations to platelets. Am J Physiol 248: H389–H395

13. Houston DS, Shepherd JT, Vanhoutte PM (1986) Aggregating human platelets cause direct contraction and endothelium-dependent relaxation in isolated canine coronary arteries. J Clin Invest 78: 539–544

14. Shimokawa H, Aazhus LL, Vanhoutte PM (1987) Porcine Coronary arteries with regenerated endothelium have a reduced endothelium-dependent respartiveness to aggregating platelets and serotonin Circulation Res 61: 256–270

15. Forstermann U, Mugge A, Alheid U, Haverich A, Frolich JC (1988) Selective attenuation of endothelium-mediated vasodilation in atherosclerotic human coronary arteries. Circ Res 62: 185–190

16. Houston DS, Vanhoutte PM (1988) Comparison of serotonergic receptor subtypes on the smooth muscle and endothelium of the canine coronary artery. J. Pharmacol Exp Ther 244: 1–10

17. Houston DS, Burnstock G, Vanhoutte PM (1987) Different P_2-purinergic receptor subtypes on endothelium and smooth muscle in canine blood vessels. J Pharmacol Exp Ther 241: 501–506

18. Flavahan NA, Shimokawa H, Vanhoutte PM (1989) Pertussis toxin inhibits endothelium-dependent relaxations to certain agonists in porcine coronary arteries. J Physiol 408: 549–560

19. Shimokawa H, Aarhus LL, Vanhoutte PM (1989) Dietary polyunsaturated fatty acids augment endothelium-dependent relaxation to bradykinin in porcine coronary microvessels. Br J Pharmacol 95: 1191–1196

20. Moncada S, Vane JR (1979) Pharmacology and endogenous roles of prostaglandin endoperoxides, thromboxane A_2 and prostacyclin. Pharmacol Rev 30: 293–331

21. Furchgott RF (1988) Studies on relaxation of rabbit aorta by sodium nitrite: The basis for the proposal that acid-activatable inhibitory factor from bovine retractor penis is inorganic nitrite and the endothelium derived relaxing factor is nitric oxide. In: Vanhoutte PM (ed) Vasodilatation: Vascular smooth muscle, peptides, autonomic nerves and endothelium. Raven, New York, pp 401–414

22. Ignarro LJ, Byrns RE, Wood KS (1988) Biochemical and pharmacological properties of endothelium-derived relaxing factor and its similarity to nitric oxide

radical. In: Vanhoutte PM (ed) Vasodilatation: Vascular smooth muscle, peptides, autonomic nerves and endothelium. Raven, New York, pp 427–436

23. Palmer RMJ, Ferrige AG, Moncada S (1987) Nitric oxide release accounts for the biological activity of endothelium-derived relaxing factor. Nature 327: 524–526

24. Palmer RMJ, Ashton DS, Moncada S (1988) Vascular endothelial cells synthesize nitric oxide from L-arginine. Nature 333: 664–666

25. Palmer RMJ, Moncada S (1989) A novel citrulline-forming enzyme implicated in the formation of nitric oxide by vascular endothelial cells. Biochem Biophys Res Commun 158: 348–352

26. Moncada S, Palmer RMJ, Higgs A (1988) The discovery of nitric oxide as the endogenous nitrovasodilator. Hypertension 12: 365–372

27. Feletou M, Vanhoutte PM (1988) Endothelium-dependent hyperpolarization of canine coronary smooth muscle. Br J Pharmacol 93: 515–524

28. Hoeffner U, Feletou M, Flavahan NA, Vanhoutte PM (1989) Canine arteries release two different endothelium-derived relaxing factors. Am J Physiol 257: H330–H333

29. Boulanger C, Hendrickson H, Lorenz RR, Vanhoutte PM (1989) Release of different relaxing factors by cultured porcine endothelial cells. Circ Res 64: 1070–1078

30. Radomski MW, Palmer RMJ, Moncada S (1987) The role of nitric oxide and cGMP in platelet adhesion to vascular endothelium. Biochem Biophys Res Commun 148: 1482–1489

31. Sneddon JM, Vane JR (1988) Endothelium-derived relaxing factor reduces platelet adhesion to bovine endothelial cells. Proc Natl Acad Sci USA 85: 2800–2804

32. Radomski MW, Palmer RMJ, Moncada S (1987) The anti-aggregating properties of vascular endothelium: Interaction between prostacyclin and nitric oxide. Br J Pharmacol 92: 639–646

33. De Mey JG, Claeys M, Vanhoutte PM (1982) Endothelium-dependent inhibitory effects of acetylcholine, adenosine diphosphate, thrombin and arachidonic acid in the canine femoral artery. J Pharm Exp Ther 222: 166–173

34. Lüscher TF, Cooke JP, Houston DS, Neves R, Vanhoutte PM (1987) Endothelium-dependent relaxations in human peripheral and renal arteries. Mayo Clin Proc 62: 601–606

35. Vanhoutte PM (1989) State of the art lecture: Endothelium and control of vascular function. Hypertension 13 (6): 658–667

36. Shimokawa H, Flavahan NA, Vanhoutte PM (1990) Natural Course of the impairment of endothelium-dependent relaxations after balloon endothelium-removal in porcine coronary arteries. Circ Res 65: 740–753

CHAPTER 7

Effects of Atherosclerosis and Regression of Lesions

D.D. HEISTAD, J.A. LOPEZ, F.M. FARACI, and M.L. ARMSTRONG

Summary. Our studies address two aspects of atherosclerosis: vasospasm, and regression of atherosclerotic lesions. One goal of the studies is to understand the pathophysiology of vasospasm. The second goal is to determine whether vasomotor abnormalities of atherosclerotic arteries are corrected by regression of lesions.

The experimental model is cynomolgus monkeys which develop atherosclerotic lesions that closely resemble those occurring in humans. Monkeys also develop moderately severe lesions in coronary, carotid, and limb arteries. We usually study blood vessels in the limb, but have also studied the coronary and cerebral circulation.

Our studies indicate that vascular responses are altered by atherosclerosis, and these changes predispose to vasospasm. The most striking and consistent change is that constrictor responses to serotonin are potentiated by atherosclerosis. There is also modest potentiation of vasoconstrictor responses to thromboxane and endothelin, and modest impairment of vasodilator responses to ADP. Our studies in vitro and several others indicate that endothelium-dependent relaxation is impaired by atherosclerosis. It is likely but not proven, that impairment of endothelium-dependent relaxation contributes to augmentation of vasoconstrictor responses in vivo.

Platelets have received considerable attention as cellular mediators of vasospasm. Vascular responses to vasoactive substances released by platelets are altered by atherosclerosis to favor vasoconstriction: vasodilator responses to ADP are impaired, and vasoconstrictor responses to serotonin and thromboxane are potentiated. Infusion of Collagen produces vasodilatation in normal and atherosclerotic monkeys, but after several minutes, large arteries of atherosclerotic monkeys constrict. These studies support the hypothesis that platelets may contribute to spasm of atherosclerotic arteries.

Atherosclerotic lesions contain many leukocytes, which are predominantly monocyte-macrophages. These cells may contribute to the formation and progression of lesions. Recently, we have suggested that vasoactive products that are released by leukocytes may produce spasm in atherosclerotic arteries. Injection of fMLP, a peptide that activates leukocytes, produces constriction of large arteries in atherosclerotic monkeys, with little effect in normal monkeys. Our studies suggest that prostaglandin E_2 may contribute to leukocyte-induced constriction of atherosclerotic arteries. Thus, activation of leukocytes, as well as platelets, may produce pronounced constriction of atherosclerotic arteries.

Department of Internal Medicine and Pharmacology, Veterans Administration Medical Center, and University of Iowa College of Medicine, Iowa City, IA 52242, USA

It is now generally accepted that vasospasm can produce myocardial ischemia, angina, and myocardial infarction. We have proposed, based on studies in monkeys, that vasospasm may also contribute to transient ischemic attacks, with amaurosis fugax, and non-occlusive mesenteric ischemia. Vasospasm may produce ischemia in situations where the degree of structural vascular obstruction is insufficient to account for ischemia.

Because plasma cholesterol concentratian can now be effectively reduced, there is strong incentive to examine hemodynamic consequences of regression of atherosclerotic lesions. Regression of lesions in monkeys is not accompanied by corresponding improvement in maximal vasodilator capacity, which is probably impaired by arterial fibrosis. In contrast, we have found that regression of atherosclerosis is accompanied by restoration of endothelium-dependent relaxation and abolition of hyperresponsiveness. We thus anticipate that vasospastic syndromes may respond to effective treatment of hypercholesterolemia.

Introduction

This review addresses two aspects of atherosclerosis: vasospasm, and regression of atherosclerotic lesions. One goal of our studies is to understand the pathophysiology and consequences of vasospasm. The second goal is to determine whether vasomotor abnormalities of atherosclerotic arteries are corrected by regression of lesions.

Until about fifteen years ago, there was little interest in vasospasm. It was argued that stiff, atherosclerotic arteries could not contract or relax, and that therefore vasospasm could not be of clinical importance. It then became apparent that vasospasm is a common clinical complication of atherosclerosis, and that it can produce ischemia and infarction [1]. Our studies have addressed several aspects of the pathophysiology of vasospasm. We have attempted to determine potential chemical mediators of spasm, cellular elements that may produce vasospasm, the role of endothelium in the propensity to vasospasm, and possible consequences of spasm.

Regression of atherosclerotic lesions was not demonstrated unequivocally until about twenty years ago. Before that, studies in rabbits failed to demonstrate regression of lesions when atherosclerotic rabbits were fed a low cholesterol diet. Studies in chickens suggested that coronary lesions regress, but aortic lesions fail to regress. Then, in a paper which has become a citation classic [2], regression of atherosclerosis was demonstrated in monkeys [3]. Now that it has become possible to effectively reduce blood cholesterol levels in humans, an urgent goal is to understand the consequences of regression of atherosclerosis. We have attempted to determine whether vasomotor abnormalities of atherosclerotic arteries are corrected by regression of lesions.

Experimental Model

The experimental model that we have used is the cynomolgus monkey. Monkeys are expensive to purchase and maintain, but three major advantages justify their cost. First, monkeys develop atherosclerotic lesions that closely resemble the

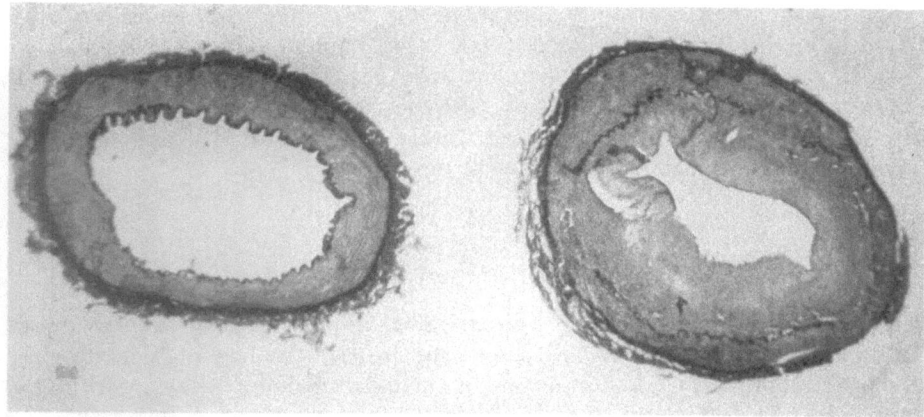

Fig. 1. Iliac artery of a normal cynomolgus monkey, and a monkey that was fed an atherogenic diet for 18 months. Marked proliferation of the intima produced a typical moderately-severe atherosclerotic lesion.

lesions that develop in humans. Second, monkeys (like humans) develop prominent atherosclerotic lesions in the coronary, carotid, limb, and other peripheral arteries (Fig. 1). No other species develops lesions of comparable severity ˙ in these arteries, unless vascular injury is superimposed on an atherogenic diet. Third, monkeys undergo regression of atherosclerotic lesions.

Before normal monkeys are studied, they are fed a diet that is low in fat (about 5%) and virtually free of cholesterol. Normal monkeys rarely have any vascular lesions. We also study hypercholesterolemic monkeys, before they develop atherosclerotic lesions. The monkeys are fed a semipurified atherogenic diet that contains about 40% fat and 0.7% cholesterol, for a period of 3–5 months. Blood cholesterol levels increase from normal levels of about 100 mg/dl to about 600 mg/dl, without an increase in triglycerides. The monkeys rarely have any atherosclerotic lesions in peripheral arteries after 3–5 months of an atherogenic diet [4].

To produce early atherosclerotic lesions, monkeys are fed the atherogenic diet for 8–10 months [5]. To produce moderately severe atherosclerosis, monkeys are usually fed the atherogenic diet for 18 months, or as long as 5 years. Atherosclerotic lesions are usually present in epicardial coronary arteries, proximal carotid arteries, and iliac arteries [6], in a distribution that closely resembles that which occurs in humans. It is our impression that more severe lesions develop in monkeys that are older when the diet is started, and less severe in young monkeys.

To study effects of regression of atherosclerosis, monkeys are fed an atherogenic diet for 18 months, and then a normal diet for 18 months. Intimal lesions are about 50% smaller than those observed in monkeys that are fed an atherogenic diet for 18 months, and virtually all of the intimal lipid is reabsorbed [7]. Regression of lesions is accompanied by fibrosis, as the density of collagen in intima is greater than that in atherosclerotic monkeys.

We induce atherosclerotic lesions by producing severe hypercholesterolemia (cholesterol level of approximately 600 mg/dl) for 18 months. Others have induced comparable lesions in monkeys with less severe hypercholesterolema (cholesterol level of approximately 350 mg/dl) for 5 years. These findings provide support for the concept that dietary cholesterol is a risk factor for development of atherosclerotic vascular disease.

Vascular Responses

Atherosclerosis predisposes to vasospasm in human [1], in atherosclerotic arteries in vitro [8], and in atherosclerotic experimental animals in vivo [4]. We have attempted to determine which chemical mediators may contribute to vasospasm. The approach that we have taken is to examine responses to a variety of possible mediators of vasospasm, and to determine whether atherosclerosis potentiates vasoconstrictor responses to these vasoactive stimuli. To account for the pronounced constriction which occurs in vasospasm, we have been looking for stimuli which are greatly potentiated by atherosclerosis.

We began our studies of vascular responsiveness in atherosclerotic monkeys in 1971. Responses to a variety of vasoconstrictor stimuli were examined, and because responses were not greatly altered, we did not believe the findings merited publication. It was not until 1980 that we published our first paper about hemodynamic consequences of atherosclerosis [9]. It would be difficult, in the current climate of funding for biomedical research, to be able to continue studies for 9 years before publication of an initial paper in an area of research. Fortunately, over the past decade our progress in these studies has been more rapid.

The major paper that led us in a productive direction was the finding by Henry [8] that contractile responses to ergonovine are greatly potentiated in the aorta of atherosclerotic rabbits, and that serotonergic mechanisms account for augmentation of the response. That observation led us to examine responsiveness to serotonin in atherosclerotic monkeys.

Serotonin produces modest constriction of large arteries in the limbs of normal monkeys in vivo. In atherosclerotic monkeys, constrictor responses of large arteries are potentiated at least ten-fold [4]. We have subsequently examined effects of serotonin in a variety of vascular beds, and the results are remarkably consistent: in coronary, cerebral, retinal, and mesenteric arteries, atherosclerosis greatly potentiates vasoconstrictor responses to serotonin [10–12]. There is also a modest potentiation of vasoconstrictor responses to thromboxane [5,13] and endothelin [14] in atherosclerotic monkeys, but the magnitude of potentiation of responses is far greater for serotonin.

We have studied vascular responses to a variety of other agonists, and have not been able to detect consistent alteration of responses in atherosclerotic monkeys. Some of these stimuli include stimulation of sympathetic nerves, norepinephrine, histamine, vasopressin, and angiotensin. Thus, there is some degree of specificity in potentiation of vasoconstrictor responses in atherosclerotic monkeys, and potentiation of responses to serotonin is particularly impressive.

Platelet Hypothesis

A major hypothesis concerning the pathophysiology of vasospasm is that platelets may initiate spasm [15,16]. The hypothesis is that platelets may adhere to atherosclerotic lesions, aggregate, and release vasoactive products.

Platelets release large amounts of preformed ADP, a potent vasodilator, and moderate amounts of serotonin. Platelets also synthesize and release thromboxane, but although thromboxane is a potent vasoconstrictor, its vasoconstrictor effects are masked by the vasodilator effect of ADP. Thus, in normal arteries, aggregation of platelets produces relaxation [17].

Our studies indicate that atherosclerosis alters vascular responses to the substances that are released by platelets. Vasodilator responses to ADP are impaired by atherosclerosis, vasodilator responses to serotonin are reversed to vasoconstriction, and vasoconstrictor responses to thromboxane are augmented (Fig. 2) [5]. Thus, vascular responses to the products that are released by platelets are altered by atherosclerosis in a direction that favors vasoconstriction, or perhaps spasm, when platelets aggregate.

Even early atherosclerotic lesions can alter vascular responses (Fig. 3) [5]. The finding is of interest because of the clinical observation that spasm may occur in arteries with only minimal atherosclerotic lesions [18].

We have performed other studies to more directly examine the effects of vasoactive products that are released by platelets. Human platelets were aggregated in vitro, and the supernatant, which contained the products released by platelets, was injected into the limbs of monkeys [19]. It was important to inject the supernatant soon after aggregation of the platelets, to minimize metabolism of thromboxane, which has a short half-life (about 30 sec). Injection of the supernatant produced vasodilatation in normal monkeys, as the vasodilator effects of ADP and serotonin must have been predominant over the vasoconstrictor effects of thromboxane. We were surprised to find that in atherosclerotic monkeys, the supernatant also produced vasodilatation. The

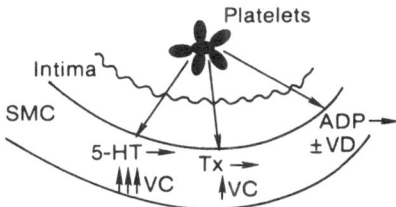

Fig. 2. The "platelet hypothesis" proposes that vasoactive products that are released when platelets aggregate may produce vasospasm. Our studies suggest that serotonin (*5-HT*) produces pronounced vasoconstrictor (*VC*) responses in atherosclerotic arteries, but not in normal arteries. Constrictor responses to thromboxane (*Tx*) are moderately potentiated by atherosclerosis. ADP does not produce vasodilatation (*VD*) of large arteries in vivo, in contrast to other studies in vitro. It is likely, however, that in normal large arteries, ADP attenuates vasoconstrictor responses to serotonin and thromboxane, and that this modulating affect of ADP is impaired in atherosclerotic large arteries

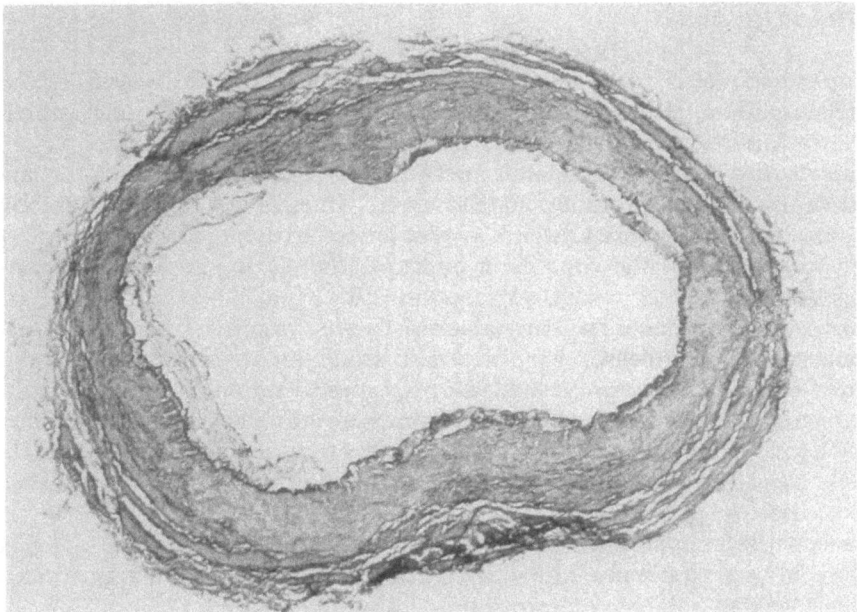

Fig. 3. Iliac artery of a monkey that was fed an atherogenic diet for 9 months. An early lesion, with intimal proliferation, is apparent on the left side of the artery

vasodilatation was less in atherosclerotic monkeys than in normal monkeys, but the response was not reversed to vasoconstriction. Thus, the vasodilator effect of ADP presumably masked the vasoconstrictor effect of serotonin and thromboxane in atherosclerotic monkeys.

In recent experiments, we have aggregated platelets in vivo by infusing purified collagen directly into the perfused limbs of monkeys [20]. Collagen produced vasodilatation in normal monkeys, as the dilator effects of ADP were predominant. In atherosclerotic monkeys collagen initially produced vasodilatation but, after several minutes, there was pronounced constriction of large arteries. The findings suggest that when platelets are activated in vivo, they may initiate constriction or perhaps spasm of atherosclerotic arteries.

Leukocyte Hypothesis

An impressive finding in many atherosclerotic arteries is that the vessel wall is infiltrated with mononuclear cells. A lot of attention has been directed towards the hypothesis that blood-borne monocytes that adhere to endothelium, or monocyte-macrophages within atherosclerotic lesions, may contribute to the atherosclerotic process [21,22] by stimulation of cell migration and proliferation.

We have proposed a new hypothesis: leukocytes may release vasoactive products that produce spasm of atherosclerotic arteries [23]. This hypothesis is

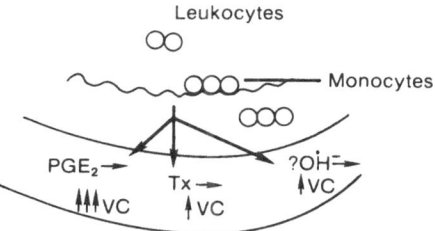

Fig. 4. The "leukocyte hypothesis" proposes that vasoactive products that are released by leukocytes may induce spasm of atherosclerotic arteries. Blood-borne leukocytes, monocytes that are adherent to endothelium, or monocyte macrophages in the arterial wall may mediate the effect. Our studies suggest that prostaglandin E_2 (PGE_2) may produce pronounced vasoconstriction (VC) in atherosclerotic arteries, thromboxane may contribute to the response, and preliminary results suggest that the hydroxyl radical (OH^-) may play an important role

based on responses to injection of a peptide, f-met-leu-phe (or fMLP), which activates leukocytes and thereby initiates release of their vasoactive products. Other investigators have examined responses to fMLP in normal animals [24], or after myocardial infarction [25], but not in atherosclerotic animals. We have found that fMLP has little effect in the limbs of normal monkeys. However, it produces profound vasoconstriction in atherosclerotic monkeys (23). The finding led us to propose that activation of leukocytes, with release of their vasoactive products, is able to initiate constriction or perhaps spasm of atherosclerotic arteries (Fig. 4). Recently, we have made similar observations in the cerebral circulation [26].

Leukocytes release a variety of vasoactive substances, including several arachidonate metabolites (thromboxane, leukotrienes, and prostaglandin E_2) and oxygen-derived free radicals. We are attempting to determine which leukocyte-derived vasoactive substances may account for constriction of atherosclerotic arteries. Our evidence to date suggests that thromboxane, but probably not leukotrienes, may contribute to leukocyte-induced constriction of atherosclerotic arteries. Prostaglandin E_2, which normally is a vasodilator, produces pronounced constriction of large arteries in atherosclerotic monkeys [23], and thus may also contribute to leukocyte-induced vasoconstriction.

Accumulations of mast cells have been observed in proximity to a coronary artery in a patient that was prone to vasospasm [27]. It is possible that histamine [28,29] or other vasoactive stubstances may account for vasospasm. This hypothesis is supported by the finding that atherosclerosis potentiates vasoconstrictor responses to histamine in coronary arteries of humans [30].

Endothelial Dysfunction

Many vasoactive stimuli, including serotonin and ADP, release endothelium-derived relaxing factor (EDRF) in normal arteries. Denudation of endothelium from normal arteries may prevent relaxation, and responses to dilator stimuli may even be reversed to contraction [31,32]. Endothelial modulation also accounts for the finding that aggregation of platelets produces relaxation in normal arteries, and contraction after removal of endothelium [17].

Endothelium is present in atherosclerotic arteries, but endothelium-dependent relaxation is greatly impaired [33–35]. Studies with a bioassay indicate that atherosclerotic arteries are able to respond normally to EDRF, and that impaired release of EDRF may account for impaired endothelium-dependent relaxation. In any case, it is quite clear that atherosclerosis impairs endothelium-dependent relaxation.

It seems logical, although perhaps not proven, that impairment of endothelium-dependent relaxation contributes to augmentation of vasoconstrictor responses in atherosclerotic arteries in vivo. For example, endothelial dysfunction may result in vasoconstrictor responses to serotonin instead of vasodilatation, or to aggregation of platelets of atherosclerotic arteries. Similarly, endothelial dysfunction may lead to vasoconstrictor responses to PGE_2, and activation of leukocytes.

Studies in patients also suggest that endothelial function is abnormal in atherosclerotic arteries [36,37]. In patients with coronary arteries that appear relatively normal, acetylcholine produces dilatation. In contrast, acetylcholine produces constriction of coronary arteries with stenotic atherosclerotic lesions [36]. Abnormal responses probably are related to endothelial dysfunction [37].

Abnormal Microcirculation

Atherosclerotic lesions are confined to large arteries. It was therefore somewhat surprising that Henry et al. [38] found that microvascular dilator responses to acetylcholine are impaired in the skeletal muscle of atherosclerotic rabbits.

We have found clear evidence that the coronary microcirculation of atherosclerotic monkeys is abnormal, even though there are no atherosclerotic lesions in the microvessels [12]. Serotonin produces vasodilatation in the coronary microcirculation of normal monkeys. In atherosclerotic monkeys, serotonin fails to produce dilatation in the coronary microcirculation in vivo. Thus, the vasomotor abnormalities in atherosclerosis occur in both diseased large arteries and in downstream coronary microvessels.

Studies in vitro also indicate that endothelial dysfunction occurs in coronary microvessels of atherosclerotic monkeys [39]. Relaxation to bradykinin, the calcium ionophore A23187, and acetylcholine is profoundly impaired by atherosclerosis. In contrast, responses to adenosine and sodium nitroprusside, which are endothelium-independent agonists, are normal. Indomethacin does not alter responses, which is evidence against the existence of a prostanoid endothelium-dependent contracting factor. The implications of this finding are clear. It is likely that the functional consequences of vasospasm are the result of vasomotor abnormalities of both the large, grossly atherosclerotic arteries and lesion-free microvessels (Fig. 5).

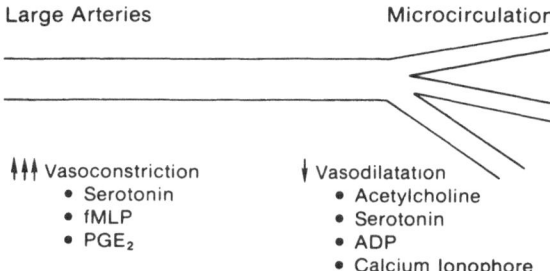

Fig. 5. Atherosclerosis produces marked potentiation of constrictor responses of large arteries to serotonin and prostaglandin E_2, and to activation of leukocytes by fMLP. In our studies responses to other vasoconstrictors are not greatly potentiated by atherosclerosis Vasodilator responses to a variety of endothelium-dependent agonists are impaired in the microcirculation of atherosclerotic monkeys

Vasospastic Syndromes

A decade ago, the possibility that vasospasm could produce myocardial ischemia was controversial. It is now generally accepted. Surprisingly, the possibility that vasospasm may produce ischemia in other organs has received little support.

For example, it is clear that platelets play a critical rold in the pathophysiology of transient ischemic attacks (TIA). Neurologists attribute most TIA to adherence of platelets to atherosclerotic lesions in the carotid arteries, with emboli to the cerebrum and retina. A role of vasospasm in the pathophysiology of TIAs is suggested by the finding that serotonin produces a large reduction in retinal blood flow in atherosclerotic monkeys, with a "flat" electroretinogram which suggests transient blindness [11]. Based on this finding, our speculation is that vasoactive substances that are released by platelets may contribute to amaurosis fugax and TIA. This hypothesis probably does not seem novel to cardiologists, who acknowledge that platelets may contribute to coronary vasospasm but the hypothesis seems novel to many neurologists.

Mesenteric ischemia sometimes occurs in the absence of vascular obstruction. The syndrome of non-occlusive mesenteric ischemia has been reported to improve after infusion of papaverine, which suggests a vasospastic origin. We infused serotonin in normal and atherosclerotic monkeys, and found enormous reductions in blood flow to the colon in atherosclerotic monkeys, but not in normal monkeys [10]. This observation led us to suggest that, if platelets aggregate at atherosclerotic plaques in the mesenteric arteries, exaggerated vasoconstrictor responses to serotonin may contribute to nonocclusive mesenteric ischemia.

It is important to emphasize that the blood levels of serotonin that we produce in our studies are probably comparable to levels that have been observed in the presence of thrombi in vivo. In several studies, we have infused serotonin into the left atrium to avoid clearance by pulmonary endothelium. We assume that 90% of serotonin is cleared in one pass through the lung, that the circulation

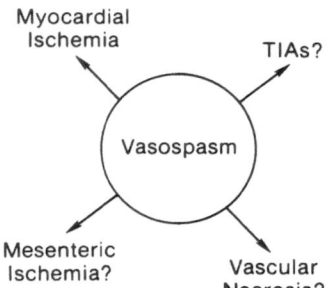

Myocardial
Ischemia

TIAs?

Vasospasm

Mesenteric
Ischemia?

Vascular
Necrosis?

Fig. 6. Vasospasm can clearly produce myocardial ischemia, with angina or myocardial infarction. We have speculated that spasm of atherosclerotic arteries may contribute to transient ischemic attacks (*TIA*), non-occlusive mesenteric ischemia, and necrosis in the wall of atherosclerotic arteries

time in monkeys is 14 sec, and that the volume of distribution is confined to plasma. We thus estimate that infusion of serotonin at 50 µm/kg/min into the left atrium produces plasma concentrations of 215 ng/ml. This plasma level is comparable to levels that have been measured downstream from an experimental platelet thrombus (213 ± 63 ng/ml) [40].

Thus, based on findings in cerebral, retinal, and mesenteric circulations [10,11], we speculate that vasospasm may contribute to ischemia in several syndromes in which the amount of structural vascular obstruction may not be sufficient to produce ischemia (Fig. 6).

Vasa Vasorum

Based on morphological observations, it has been known for many years that vasa vasorum proliferate in atherosclerotic arteries [41]. We have used microspheres to obtain the first measurements of blood flow through vasa. There is a remarkable increase in flow through vasa in the aorta, coronary arteries, and carotid arteries of atherosclerotic monkeys [42–44].

Vasa vasorum in atherosclerotic arteries are hyperresponsive to serotonin [44,45]. This hyperresponsiveness may have functional implications. When serotonin produces constriction of atherosclerotic coronary and carotid arteries [11,12], it is likely that the metabolic needs of the arteries also increase. The combination of increased metabolic needs of the vessel wall, combined with reduction in blood flow through vasa [44,45], may predispose the arterial wall to ischemia. Thus, the pathophysiology of vascular necrosis, which characterizes atherosclerotic arteries, may be related in part to hyperresponsiveness of vasa vasorum to constrictor stimuli.

We have obtained evidence that blood flow to vasa vasorum in coronary arteries, which increases almost tenfold in atherosclerotic monkeys, returns to virtually normal levels after regression of lesions [45]. It has been proposed that vasa may play an important role in complications of atherosclerosis, because vasa may be prone to rupture, with production of intimal hemorrhage, and because vasa may deliver vasoactive substances to the arterial wall and thereby predispose to vasospasm. If these concepts are valid, loss of vasa during regression of atherosclerosis [45] may have important beneficial consequences.

Regression of Atherosclerosis

Now that physicians can effectively reduce the concentration of plasma cholesterol in hypercholesterolemic patients, there is a compelling reason to determine whether reduction of cholesterol is beneficial. Epidemiological evidence provides support for the concept that treatment of hypercholesterolemia is beneficial [46]. Very little is known, however, about hemodynamic consequences of reduction of hypercholesterolemia.

Morphological evidence of regression of atherosclerosis, with reabsorption of lipid from the intima [3,7], is not always accompanied by corresponding improvement in maximal vasodilator capacity [7]. It is likely that arteries undergo fibrosis during regression, which restricts improvement in vasodilator capacity. Exercise-induced angina pectoris and intermittent claudication are produced by impairment of maximal vasodilator capacity. In the light of persistent impairment of maximal vasodilator capacity with regression of atherosclerosis [7], we are not optimistic that exercise-induced angina or intermittent claudication will consistently improve with effective treatment of hypercholesterolemia. We have, however, found that regression of atherosclerosis restores endothelium-dependent relaxation to normal [47], and abolishes hyperresponsiveness of arteries [48,49]. Thus, we are optimistic that vasospastic syndromes may improve during regression of atherosclerosis. It seems likely, therefore, that vasospastic angina and TIA may respond to effective treatment of hypercholesterolemia. A study of patients who were treated for familial hypercholesterolemia [50] provides some support for this hypothesis.

Acknowledgments. We thank Ms. Marge Keaough for typing the manuscript. The original studies described in the manuscript were supported by a Medical Investigator Award, Associate Investigator Award, and research funds from the Veterans Administration, and by National Institutes of Health grants HL-14230, NS 24621, HL 14388, and HL 16066.

References

1. Maseri A, L'Abbate A, Barold G, Chierchia S, Marzilli M, Ballestro AM, Severi S, Parodi O, Biagini A, Distante A, Pesola A (1978) Coronary vasospasm as a possible cause of myocardial infarction. N Eng J Med 299: 1271–1277
2. Armstrong M (1986) Citation Classic. Current Contents 29: 16
3. Armstrong M, Warner E, Connor W (1970) Regression of coronary atheromatosis in rhesus monkeys. Circ Res 27: 59–67
4. Heistad D, Armstrong M, Marcus M, Piegors DJ, Mark AL (1984) Augmented responses to vasoconstrictor stimuli in hypercholesterolemic and atherosclerotic monkeys. Circ Res 54: 711–718
5. Lopez JAG, Armstrong M, Piegors D, Heistad DD (1989) Effect of early and advanced atherosclerosis on vascular responses to serotonin, thromboxane A_2, and ADP. Circulation 79: 698–705
6. Armstrong M. Warner E (1971) Morphology and distribution of diet-induced atherosclerosis in rhesus monkeys. Arch Pathol. 92: 395–401

7. Armstrong M, Heistad D, Marcus M, Piegors DJ, Abbovd FM (1983) Hemodynamic sequelae of regression of experimental atherosclerosis. J Clin Invest 71: 104–2113

8. Henry PD, Yokoyama M (1980) Supersensitivity of atherosclerotic rabbit aorta to ergonovine: Mediation by a serotonergic mechanism. J Clin Invest 66: 306–313

9. Heistad DD, Marcus ML, Piegors DJ, Armstrong ML (1980) Regulation of cerebral blood flow in atherosclerotic monkeys. Am J Physiol (Heart Circ Physiol 8): H539–H544

10. Lopez JAG, Brown BP, Armstrong ML, Piegors DJ, Heistad DD (1989) Response of the mesenteric circulation to serotonin in normal and atherosclerotic monkeys: Implications for the pathogenesis of nonocclusive intestinal ischemia. Cardiovasc Res 23: 117–124

11. Williams JK, Baumbach GL, Armstrong ML, Heistad DD (1989) Hypothesis: Vasoconstriction contributes to amaurosis fugax. J Cereb Blood Flow Metab 9: 111–116

12. Chilian WM, Dellsperger KC, Layne SM, Eastham CL, Armstrong MA, Marcus ML, Heistad CD (1990) Effects of atherosclerosis on the coronary circulation: Potentiation of arterial and microvascular constrictor responses. Am J Physiol (Heart 27) H529–539

13. Faraci FM, Williams JK, Breese KR, Armstrong ML, Heistad DD (1989) Atherosclerosis potentiates constrictor responses of cerebral and ocular blood vessels to thromboxane. Stroke 20: 242–247

14. Lopez JAG, Armstrong ML, Piegors DJ, Heistad DD (1989) Vascular responses to endothelin in atherosclerotic primates (abstract). Circulation (Suppl II) 80: II586

15. Vanhoutte PM, Houston DS (1985) Platelets, endothelium, and vasospasm. Circulation 72: 728–734

16. Willerson JT, Hillis DH, Winniford MW, Buja LM (1986) Speculation regarding mechanisms responsible for acute ischemic heart disease syndromes. J Am Coll Cardiol 8: 245–250

17. Houston DS, Shepherd JT, Vanhoutte PM (1986) Aggregating human platelets cause direct contraction and endothelium-dependent relaxation of isolated canine coronary arteries. J Clin Invest 78: 539–544

18. Heupler FA Jr (1980) Syndrome of symptomatic coronary arterial spasm with nearly normal coronary arteriograms. Am J Cardiol 45: 873–881

19. Lopez JAG, Brotherton AFA, Armstrong ML, Piegors DJ, Heistad DD (1988) Vascular responses to activated platelets in atherosclerotic primates (abstract) FASEB J 2: A943

20. Lopez JAG, Hoak JC, Armstrong ML, Piegors DJ, Heistad DD (to be published) Effect of intravascular collagen in atherosclerotic primates (abstract. J Am Coll Cardiol

21. Gerrity RG (1981) The role of the monocyte in atherogenesis. Transition of blood-borne monocytes into foam cells in fatty lesions. Am J Pathol 103: 181–190

22. Faggiotto A, Ross R, Harker L (1984) Studies of hypercholesterolemia in the nonhuman primate. Changes that lead to fatty streak formation. Arteriosclerosis 4: 323–340

23. Lopez JAG, Armstrong ML, Harrison DG, Piegors DJ, Heistad DD (1989) Vascular responses to leukocyte products in atherosclerotic primates. Circ Res 65: 1078–1086

24. Gillespie MN, Booth DC, Friedman BJ, Cunningham MR, Jay M, DeMaria AN (1988) fMLP provides coronary vasoconstriction and myocardial ischemia in rabbits. Am J Physiol 254 (Heart Circ Phusiol 23): H481–H486

25. Evers AS. Murphree S, Saffitz JE, Jakschik BA, Needleman P (1985) Effects of endogenously produced leukotrienes, thromboxane, prostaglandins on coronary vascular resistance in rabbit myocardial infarction. J Clin Invest 75: 992–999

26. Lopez JAG, Faraci FM, Breese KR, Armstrong ML, Heistad DD Response of cerebral and ocular vessels to activated leukocytes in atherosclerotic primates (abstract). Clin Res 38: 292A, 1990

27. Forman MB, Oates JA, Robertson D, Robertson RM, Roberts LJ II, Virmani R (1985) Increased adventitial mast cells in a patient with coronary spasm. N Engl J Med 313: 1138–1141

28. Shimokawa H, Tomoike H, Nabeyama S, Yamamoto H, Araki H, Namamura M (1983) Coronary artery spasm induced in atherosclerotic miniature swine. Science 221: 560–562

29. Yamamoto Y, Tomoike H, Egashira K, Nakamura M (1987) Attenuation of endothelium-related relaxation and enhanced responsiveness of vascular smooth muscle to histamine in spastic coronary arterial segments from miniature pigs. Circ Res 61: 772–778

30. Kalsner S, Richards R (1984) Coronary arteries of cardiac patients are hyperreactive and contain stores of amines: A mechanism for coronary spasm. Science 223: 1435–1437

31. Furchgott F R, Zawadzki J V (1980) The obligatory role of endothelial cells in the relaxation of arterial smooth muscle by acetylcholine. Nature 288: 373–376

32. Furchgott RF (1983) Role of the endothelium in responses of vascular smooth muscle. Circ Res 53: 557–573

33. Verbeuren TJ, Jordaens FH, Zonnekeyn LL, VanHobe CE, Coene M-C, Herman AG (1986) Effect of hypercholesterolemia on vascular reactivity in the rabbit. Circ Res 58: 552–564

34. Habib J, Bossaller C, Wells S, Williams C, Morrisett JD, and Henry PD (1986) Preservation of endothelium-dependent vascular relaxation in cholesterol-fed rebbit by treatment with the calcium blocker PN 200110. Circ Res 58: 305–309

35. Freiman PC, Mitchell GC, Heistad DD, Armstrong ML, Harrison DG (1986) Atherosclerosis impairs endothelium-dependent vascular relaxation to acetylcholine and thrombin in primates. Circ Res 58: 783–789

36. Ludmer PL, Selwyn AP, Shook TL, Wayne RR, Mudge GH, Alexander RW, Ganz P (1986) Paradoxical vasoconstriction induced by acetylcholine in atherosclerotic coronary arteries. N Engl J Med 315: 1046–1051

37. Förstermann U, Mögge A, Alheid U, Haverich A, Frölich JC (1988) Selective attenuation of endothelium-mediated vasodilation in atherosclerotic human coronary arteries. Circ Res 62: 185–190

38. Yamamoto H, Bossaller C, Cartwright J Jr, Henry PD (1988) Videomicroscopic demonstration of defective cholinergic arteriolar vasodilation in atherosclerotic rabbit. J Clin Invest 81: 1752–1758

39. Selke FW, Armstrong ML, Harrison DG Endothelium-dependent vascular relaxation is abnormal in the coronary microcirculation of atherosclerotic primates. Circulation 81: 1586–1593, 1990

40. Benedict CR, Mathew B, Rex KA, Cartwright J Jr, Sordahl LA (1986) Correlation of plasma serotonin changes with platelet aggregation in an in vivo dog model of spontaneous occlusive coronary thrombus formation. Circ Res 58: 58–67

41. Geiringer E (1951) Intimal vascularization and atherosclerosis. J Path Bact 63: 201–211

42. Heistad DD, Armstrong ML, Marcus ML (1981) Hyperemia of the aortic wall in

atherosclerotic monkeys. Circ Res 48: 669–675

43. Heistad DD, Armstrong ML (1986) Blood flow through vasa vasorum of coronary arteries in atherosclerotic monkeys. Arteriosclerosis 6: 326–331
44. Williams JK, Orgren KI, Armstrong ML, Heistad DD (1989) Vasa vasorum in the carotid sinus of atherosclerotic monkeys: Implications for baroreceptor function. Atherosclerosis 78: 25–32
45. Williams JK, Armstrong ML, Heistad DD (1988) Vasa Vasorum in atherosclerotic coronary arteries: Responses to vasoactive stimuli and dietary treatment. Circ Res 62: 515–523
46. The lipid research clinics coronary primary prevention trial results. (1984) JAMA 251: 351–356
47. Harrison DG, Armstrong ML, Freiman PC, Heistad DD (1987) Restoration of endothelium-dependent relaxation by dietary treatment of atherosclerosis. J Clin Invest 80: 1808–1811
48. Heistad DD, Mark AL, Marcus ML, Piegors DJ, Armstrong ML (1987) Dietary treatment of atherosclerosis abolishes hyperresponsiveness to serotonin: Implications for vasospasm. Circ Res 61: 346–351
49. Heistad DD, Breese K, Armstrong ML (1987) Cerebral vasoconstrictor responses to serotonin after dietary treatment of atherosclerosis. Stroke 18: 1068–1073
50. Kuo P, Toole J, Schaaf J, Jones A, Wilson AC, Kostis JB, and Moreyra AE (1987) Extracranial carotid arterial disease in patients with familial hypercholesterolemia and coronary artery disease treated with colestipol and nicotinic acid. Stroke 18: 716–721

CHAPTER 8

Vascular Regulation in Atherosclerosis

P.D. Henry[1]

Summary. Physiologically, requirements for increased organ perfusion are met acutely by vasodilation, and chronically by the growth of pre-existent and new vascular channels. Current evidence indicates that the arteriosclerotic circulation undergoes several alterations that limit acute and chronic hyperemic responses. Macrovascular occlusive disease (atheroma, thrombosis) is not the only factor that may limit flow. Vasomotor tone, the sum of dilating and constricting influences, may be altered in favor of constriction, both at the macrovascular and microvascular levels. Major mechanisms that have been demonstrated include sensitization of arterial smooth muscle to specific endogenous constrictors (histamine, serotonin), and impaired endothelium-dependent vasodilation. The latter appears to affect particularly the arterial microcirculation which importantly determines organ perfusion to a large extent. Chronic adjustments include increases in macrovascular arterial diameter, and the generation of new arterial pathways (collateralization, microvasularization, angiogenesis). Recent experiments in our laboratory have demonstrated that arterial growth evoked by chronic increases in arterial flow is dependent on the presence of a normally functioning endothelium. After mechanical de-endothelialization of large arteries, or in the presence of hypercholesterolemic endothelial injury, vascular growth is impaired. Ongoing experiments focusing on the microcirculation further indicate that the growth of new small vessels (angiogenesis) may likewise be impaired in the presence of hypercholesterolemia. The results suggest that atherosclerosis may affect acute and chronic flow adjustments by multiple mechanisms. It is hoped that an understanding of these mechanisms will provide a new basis for the treatment of ischemic syndromes.

Introduction

During the past 15 years, evidence has been provided that atherosclerosis produces complex changes in vascular regulation [1] Although myocardial ischemia has often been studied by acutely constricting single coronary arteries in animals without atherosclerosis, it is clear that such models do not mimic the

---

[1] Department of Medicine, Molecular Physiology and Biophysics, Baylor College of Medicine, Houston, TX77030,USA

pathophysiology of thrombotic coronary occlusions in the presence of coronary disease. In this paper, we briefly consider processes that aggravate or improve coronary flow reserve in the presence of progressive coronary disease.

Pro-Ischemic Factors (Tables 1 and 2)

In the past, episodes of myocardial ischemia have often been ascribed to flow limitations imposed by fixed coronary obstructions [1]. This clinical interpretation may seem surprising considering that the rigid tube hypothesis provides no plausible explanation for variable episodes of ischemia at rest. Modern coronary arteriography and fibrinolytic therapy have established that coronary thrombosis

Table 1. Pro-ischemic factors

Occlusive disease
Atheromatous, thrombotic, embolic
Defective arterial growth (angiogenesis)
Defective luminal expansion of arteries
Defective formation of new channels (collaterals)
Blood hyperviscosity
Hyperaggregability of red blood cells
Hyperaggregability of platelets
Hypercoagulability of plasma (coagulation factors and products)
Increased constrictor tone
Supersensitivity of vascular smooth muscle to vasoconstrictors
  (histamine, serotonin, endothelin)
Increased release of constrictors (thromboxane)
Decreased vasodilator tone
Insensitivity to endothelium-dependent vasodilators
Decreased release of vasodilators (EDRF, PGI_2)

Table 2. Acute coronary syndromes

Non-thrombotic mechanisms
Spontaneous coronary spasm
Drug-induced coronary spasm
– Ergotism
– Withdrawal from nitroglycerin exposure
– Cocaine
Intramural hemorrhage
– Coronary dissection
– Rupture of vasa vasorum
Coronary embolism
– Endocardiac thrombi (myocardial disease, infective endocarditis)
– Valvular calcific disease
Thrombotic mechanisms
Atheroma rupture ("atherothrombosis")
Arterial trauma (cardiac contusion)
Arteritis (auto-immune disease, transplant allosensitization)

is the most common mechanism of acute coronary syndromes. Pathological studies suggest that thromboses occur usually at the site of a rupturing or fissuring plaque [2]. It has been suggested that lipid-rich foam cell lesions are particularly prone to rupture [3]. In support of this hypothesis are controlled arteriographic studies [4] which show that complete coronary occlusions frequently occur in arteries with minimal or no stenosis. It should be pointed out, however, that complicated coronary lesions may occur in the absence of coronary stenosis [5,6] Therefore, minimal stenoses should not be equated with early (small) foam cell lesions. In addition, early foam-cell rich lesions in childhood [5,7] and in experimental animals are not associated with coronary thrombosis and myocardial infarction. Inhomogeneity was mistakenly edited to non-homogeneity. Under the influence of intrinsic vasomotor stresses and of pulsatile or bending movements generated by the action of the heart, hard plaque components could cut into soft components by mechanisms resembling those of tissue injury after compound bone fractures. Plaque rupture is thought to trigger thrombosis by exposure of blood to thrombogenic materials such as collagen or phospholipids. Advanced atheromas are often surrounded by a dense network of vasa vasorum [6]. Microvessels, like foam cells, may be a plaque component susceptible to mechanical or chemical injury. Intraplaque hemorrhages appear to be a plausible cause for plaque rupture and intraluminal thrombosis. Autopsy studies by Maher and Cuénoud [8] suggest that intramural hemorrhages could be responsible for thrombotic plaque growth in the absence of luminal thrombosis.

Chronic increases in arterial flow as those produced by a downstream AV-fistula are known to result in luminal expansion of the high-flow arterial segments [9] During partial atheromatous occlusions of arteries, their narrowed lumens are similarly exposed to high flow. One important question is whether such hemodynamic changes evoke an arterial expansion capable of maintaining arterial patency despite the presence of space-occupying lesions. Glagov et al. [10] have suggested that coronary arteries undergo a compensatory enlargement as long as major portions of the wall remain free of lesions. We have demonstrated that high arterial flow triggers an arterial expansion (growth) only in the presence of an intact endothelium [11] In rabbits with hypercholesterolemia, carotid arteries chronically perfused at high flow rates exhibit a limited growth response. In further studies, we have shown that hypercholesterolemia in rabbits affects arterial growth also at the microvascular level (unpublished observations).

One major consideration is to determine whether atherosclerosis is associated with an impaired potential for forming collaterals. In dogs, acute circumflex occlusion produces an infarction of the posterior papillary muscle within 30 min. However, if the complete occlusion is produced slowly over a 1- to 2-week period, no infarction occurs and coronary flow reserve to the papillary muscle is preserved [12]. Such phenomena are almost certainly due to the development of collaterals of a rate fast enough to keep pace with the progression of occlusive disease. The important question is whether such slow occlusions would be as well tolerated in the presence of an atherogenic dyslipidemia (hypercholesterolemia). In addition to limited formation of new arterial channels, established collaterals are threatened by the progression of atherosclerosis. Atheroma formation tends to involve so-called stagnation points localized at arterial branch points (flow dividers) [13]. Therefore, the distribution of atheromatous plaques has a tendency to obstruct the takeoff of collaterals.

One area where the simple coronary occlusion model fails completely is in mimicking hematologic changes associated with dyslipidemia (hypercholesterolemia). It has been shown that hypercholesterolemia produces a variety of procoagulant changes, including hyperaggregability of platelets [14] and red blood cells [15]. Hyperaggregability of red blood cells, a major mechanism of blood hyperviscosity, appears to correlate well with coronary risk [15]. In addition, numerous blood clotting factors such as fibrinogen levels correlate with coronary risk [16]. Other plasma components contained in hypercholesterolemic blood could play a pivotal role in vascular regulation. In particular, lysophospholipids may determine alterations in the cellular behavior and interactions of white cells, red cells, and endothelial cells [17,18]. Despite overwhelming evidence that dyslipidemia is associated with circulating factors which tend to affect vascular regulation, these factors often receive little consideration in studies on myocardial ischemia.

We have previously demonstrated that hypercholesterolemia in rabbits is associated with a supersensitivity to the constrictor effects of selected monoamines [19]. In recent years, many agonists acting on smooth muscle as direct constrictors have been shown to promote also the release of vasoldilator factors (EDRF) from endothelial cells. Since endothelium-dependent relaxation is impaired in hypercholesterolemic states, monoaminergic constrictive hyperreactivity has been ascribed to decreased dilator tone. We have, however, observed monoaminergic supersensitivity in rabbit aortic medial strips after removal of the endothelium. Monoaminergic endothelium-independent supersensitivity has been also observed in Watanabe rabbits [20] and dogs [21]. It would be important to determine whether supersensitivity of vascular smooth muscle to monoamines (histamine, serotonin, ergonovine) involves arteriolar smooth muscle, which plays an important role in regulating organ perfusion. Recently, atherosclerotic arteries have also been shown to be hyperreactive to endothelin [22].

During the past ten years, numerous studies have underscored the importance of the endothelium in regulating arterial tone. We have demonstrated that arteries isolated from hypercholesterolemic animals, or from patients with coronary disease, exhibit a defective endothelium-dependent relaxation which is usually most marked for acetylcholine, but involves other endothelium-dependent agonists such as ATP, bradykinin, and substance P [23,24]. In addition, we have demonstrated that the endothelial defect also manifests itself at the level of arterioles (20–30 μm diameter) in vivo [25]. Of considerable interest are recent reports by Yasue et al. [26] and Vita et al. [27] which provide evidence that hypercholesterolemia in patients without angiographically demonstrable coronary disease is, like in animals, associated with defective endothelium-dependent coronary dilation. The pathophysiological significance of these

Table 3. Anti-ischemic factors (Compensatory mechanisms against ischemia)

Recanalization
Activation of anti-atherogenic mechanisms (reverse cholesterol transport)
Thrombolysis
Arterial growth (angiogenesis)
Arterial luminal expansion
Neovascularization (collateral formation)
Decreased constrictor tone
Supersensitivity of vascular smooth muscle to vasodilators (EDRF)
Decreased release of constrictors (endothelin)
Increased dilator tone
Supersensitivity of vascular smooth muscle to vasodilators (EDRF)
Increased release of vasodilators (histamine, bradykinin)

findings remains unclear, but it is reasonable to assume that defective endothelial function in man may limit coronary dilator reserve. An impaired production of prostacyclin by atherosclerotic endothelium may also contribute to a limited dilator reserve [28].

Anti-Ischemic Factors (Table 3)

The severity of occlusive coronary disease can improve over time independently of changes in coronary risk factors that are causally related to the disease. Major factors include a remodeling of arteries with or without reduction in atheroma size. In the face of persistent hyperlipidemia, the major mechanism contributing to the halting of the progression of occlusive arterial disease is removal of excess lip (cholesterol) from the arterial wall (reverse cholesterol transport) and lysis of thrombotic deposits [29,30]. As already mentioned, arteries may undergo a growth response to limit the impingement of atherosclerotic lesions on luminal patency. Of considerable interest is the observation that receptor-pharmacological changes that tend to augment constrictor responses may in part be counteracted by changes favoring vasodilation. For instance, there is evidence that atherosclerotic smooth muscle is sensitized to EDRF and nitrovasodilators, perhaps because defective production of EDRF by atherosclerotic endothelium determines an up-regulation of EDRF (NO)-dependent smooth muscle relaxation.

Conclusion

It has become clear that coronary atherosclerosis may influence myocardial perfusion by multiple mechanisms. In focussing on mechanisms of myocardial ischemia, it will be important to pay attention to atherosclerosis itself. In previous trials aimed at comparing medical versus surgical therapy of coronary disease, medical therapy included no treatment for the disease from which the patients were suffering--atherosclerosis [31].

References

1. Henry PD (1989) Inappropriate coronary vasomotion excessive constriction and insufficient dilation. In: Sperelakis N (ed) Physiology and pathophysiology of the heart. Kluwer Academic, pp 975-991
2. Davies MJ, Thomas AC (1985) Plaque fissuring: The cause of acute myocardial infarction, sudden ischemic death, and crescendo angina. B Heart J 53: 363-373
3. Fuster V, Badimon L, Cohen M, Ambrose JA, Badimon JJ, Chesebro J (1988) Insights into the pathogenesis of acute ischemic syndromes. Circulation 77: 1213-1220
4. Lichtlen PR, Hugenholtz PG, Rafflenbeul W, Hecker H, Jost S, Deckers JW (to be published) Retardation of coronary artery disease in man by the calcium channel blocker nifedipine; results of international nifedipine trial on antiatherosclerotic therapy (INTACT). Lancet 335: 1109-1113, 1990
5. Stary HC (1983) Macrophages in coronary artery and aortic intima and in atherosclerotic lesions of children and young adults up to age 29. In: Schettler G, Gotto AM, Middelhoff G, Habenicht AJR, Jurutka KR (eds) Atherosclerosis VI. Springer, Berlin Heidelberg New York, pp 462-466
6. Kamat BR, Galli SJ, Barger AC, Lainey LL, Silverman KJ (1987) Neovascularization and coronary histologic analysis. Hum Pathol 18: 1036-1042
7. Stary HC (1989) Evolution and progression of atherosclerotic lesions in coronary arteries of children and young adults. Arteriosclerosis 9: I-19-I-32
8. Maher M, Cuénoud HF (1989) Neovascularization, hemorrhages and iron deposits in coronary arteries after recent myocardial infarction. FASEB J 3: A1218
9. Kamiya A, Togawa T (1980) Adaptive regulation of wall shear stress to flow change in the canine carotid artery. Am J Physiol 239: H14-H21
10. Glagov S, Weisenberg E, Zarins CK, Stankunavicius R, Kolettis GJ (1987) Compensatory enlargement of human atherosclerotic coronary arteries. N Engl J Med 316: 1371-1375
11. Yamamoto H, Sartori M, Cartwright J, Henry PD (1987) Carotid artery expansion after opposite carotid occlusion in rabbits: suppresion by deendothelialization and hypercholesterolemia. Circulation 76: 55
12. Lambert PR, Hess DS, Bache RJ (1977) Effect of exercise on perfusion of collateral-dependent myocardium in dogs with chronic coronary artery occlusion. J Clin Invest 59: 1-7
13. Stehbens WE (1983) Fluid dynamic approaches to atherosclerosis. In: Schettler G, Nerem RM, Schmid-Schonbein H, Morl H, Diehm C (eds) Fluid dynamics as a localizing factor for atherosclerosis. Springer, Berlin Heidelberg New York Tokyo, pp 3-8
14. Insel PA, Nirenberg P, Turnbull AJ, Shattil SJ (1978) Relationships between membrane cholesterol, α-adrenergic receptors, and platelet function. Biochemistry 17: 5269-5274
15. Ruhenstroth-Bauer G, Porz P, Boss N, Lehmacher W, Stamm D (1985) The erythrocyte aggregation value as a measure of the risk of myocardial infarction and arteriosclerosis of peripheral arteries. Clin Cardiol 8: 529-534
16. DiMinno G, Mancini M (1990) Measuring plasma fibrinogen to predict stroke and myocardial infarction. Arteriosclerosis 10: 1-7
17. Kugiyama K, Kerns SA, Morrisett JD, Roberts R, Henry PD (1990) Impairment of endothelium-dependent arterial relaxation by lysolecithin in modified low-denisty lipoproteins. Nature 344: 160-162

18. Kearns MW, Rauch AL (1988) Lysophosphophatidylcholine circulates in plasma at levels that are toxic. FASEB J 2: A969
19. Henry PD, Yokoyama M (1980) Supersensitivity of atherosclerotic rabbit aorta to ergonovine. J Clin Invest 66: 306–313
20. Haugh CJ, Atkinson JB, Bluth RF, Swift LL, Robertson RM (1988) Genetically-induced atherosclerosis augments vascular smooth muscle contraction. Arteriosclerosis 8: 623a
21. Kawachi Y, Tomoike H, Maruoka Y, Kikuchi Y, Araki H, Ishii Y, Tanaka K, Nakamura M (1984) Selective hypercontraction caused by ergonovine in the canine coronary artery under conditions of induced atherosclerosis. Circulation 69: 441–450
22. Lopez JAG, Armstrong ML, Piegors DJ, Heistad DD (1989) Vascular responses to endothelin in atherosclerotic primates. Clin Res 37: 883A
23. Habib JB, Bossaller C, Wells S, Williams C, Morrisett JD, Henry PD (1986) Preservation of endothelium-dependent vascular relaxation in cholesterol-fed rabbit by treatment with the calcium blocker PN 200110. Circ Res 58: 305–309
24. Bossaller C, Habib GB, Yamamoto H, Williams C, Wells S, Henry PD (1987) Impaired muscarinic endothelium-dependent relaxation and cyclic quanosine 5'-monophosphate formation in atherosclerotic human coronary artery and rabbit aorta. J Clin Invest 79: 170–174
25. Yamamoto H, Bossaller C, Cartwright Jr. J, Henry PD (1988) Videomicroscopic demonstration of defective cholinergic arteriolar vasodilation in atherosclerotic rabbit. J Clin Invest 81: 1752–1758
26. Yasue H, Matsuyama K, Matsuyama K, Okumura K, Morikami Y, Ogawa H (1990) Responses of angiographically normal human coronary arteries to intracoronary injection of acetylcholine by age and segment: Possible risk of early coronary atherosclerosis. Circulation 81: 482–490
27. Vita JA, Treasure CB, Nabel EG, McLenachan JM, Fish RD, Yeung AC, Vekshtein VI, Selwyn AP, Ganz P (1990) Coronary vasomotor response to acetylcholine relates to risk factors for coronary artery disease. Circulation 81: 491–497
28. Pomerantz KB, Hajjar DP (1989) Eicosanoids in regulation of arterial smooth muscle cell phenotype, proliferative capacity, and cholesterol metabolism. Arteriosclerosis 9: 413–429
29. Reichl D, Miller NE (1989) Pathophysiology of reverse cholesterol transport. Insights from inherited disorders of lipoprotein metabolism. Arteriosclerosis 9: 785–797
30. Marder VJ, Sherry S (1988) Thrombolytic Therapy: Current status. New Engl. J Med. 318: 1512–1520; 1585–1595
31. CASS Principal Investigators (1983) Coronary artery surgery strdy (CASS): A randomized trial of coronary artery bypass surgery. Circulation 68: 939–950

CHAPTER 9

Characteristics of Vascular Smooth Muscle Cell Membranes and Their Modifying Factors

K. Kitamura, T. Itoh, Y. Ito, and H. Kuriyama[1]

Summary. In vascular smooth muscle, there are K channels (Ca-sensitive and insensitive, ATP-sensitive and less sensitive), Na channels (tetrodotoxin-sensitive and less sensitive), voltage-sensitive Ca channels, and receptor-activated (operated) cation-channels. These channel activities are important in the maintenance of electrical events in the smooth muscle cell membrane and in Ca regulation. These ionic channels are regulated directly or indirectly via synthesis of second messengers by various factors such as adrenergic innervation, humoral substances, endothelium-derived contracting and relaxing factors (endothelium-derived relaxing and hyperpolarizing factors, prostaglandin I_2, endothelin, and thromboxane A_2-like substance). In this article, the biophysical features of vascular smooth muscle cell membranes are reviewed in relation to coronary vasospasm, as well as the actions of some clinical drugs—nitro-(nitroso-, nitrite-) compounds, β-adrenoceptor blockers, Ca antagonists, and K channel openers—on ionic channels in vascular smooth muscle cells.

Introduction

The regulation of blood pressure is achieved mainly via changes in cardiac output, together with changes in the resistance and capacitance of the vascular beds. In the maintenance of homeostasis, the blood flow in the living body is regulated systemically by various factors such as humoral substances, neurotransmitters, autacoids, and bioactive peptides. Furthermore, smooth muscle cells themselves possess a property that contributes to the regulation of vascular tone via regulation of the ionic channel, and Ca regulation at the sarcoplasmic reticulum. Endothelial cells also regulate vascular tone by release of substances (factors) which induce either vasodilatation or vasoconstriction.

Cardiac muscles have a high oxygen consumption, and as a consequence, the amount of oxygen supplied through the coronary blood flow plays an essential role in the physiological function of cardiac muscle cells. Thus, organic and

---

[1] Department of Pharmacology, Faculty of Medicine, Kyushu University, Fukuoka, 812 Japan

functional stenosis of the coronary artery may induce severe cardiac failure, as in angina pectoris. The coronary arteries of several species (rat, rabbit, cat, dog, pig) show remarkable regional and species differences in the characteristics of the smooth muscle membranes. In this article, we review recent investigations made on the biophysical properties of vascular smooth muscle cell membranes, and on the factors influencing them. Special attention is focussed on the features of the smooth muscle cells of coronary arteries.

General features of electrical and mechanical responses of vascular smooth muscle cells

Properties of Vascular Smooth Muscle

The contractile proteins of smooth muscle cells (including vascular smooth muscle) possess specific features that differ from those of skeletal and cardiac muscle cells.

1. The Ca binding protein in skeletal and cardiac muscles is troponin C (actin-linked modulation theory). However, in smooth muscles, the presence of leiotonin (instead of troponin C) has been suggested as the Ca-binding protein [1–5], although the Ca-binding protein for triggering contraction is thought to be calmodulin, i.e., a complex formation of 4Ca-calmodulin-myosin light chain kinase-myosin light chain (20kd proteins) may lead to contraction (myosin-linked modulation or phosphorylation theory) [6–11].
2. In visceral smooth muscle, two folding sites (6S and 10S) of H-meromyosin are found (10S is a specific site for smooth muscles), and conformational changes between 6S and 10S have a close relation to bridge formation between actin and myosin. By contrast, skeletal and cardiac muscles possess only one folding site [12,13].
3. In skeletal and cardiac muscle, superprecipitation resembling contraction occurs by interaction of actin and myosin with tropomyosin, and is inhibited by addition of the troponin complex. Addition of Ca causes the superprecipitation to recur. In smooth muscle, superprecipitation does not occur in the presence of actin and myosin, but addition of the Ca-calmodulin complex does produce superprecipitation [2,3].
4. In smooth muscle cells, okadaic-acid-sensitive myosin phosphatase seems to play an essential role in relaxation [14], but in skeletal and cardiac muscles, the role of this phosphatase has not been clarified.

Figure 1 represents a schematic path for myosin light chain phosphorylation following Ca-calmodulin binding (A), and the schematic arrangement of the site of phosphorylation of the myosin head and the myosin light chain in relation to actin (B).

Differences between smooth muscle cells and cardiac or skeletal muscle cells occur not only in their mechanical properties, but also in their morphological structure; in cardiac and skeletal muscle, the transverse structure and sarcoplasmic reticulum (SR) forms a triad or diad, and the presence of fine foot-like

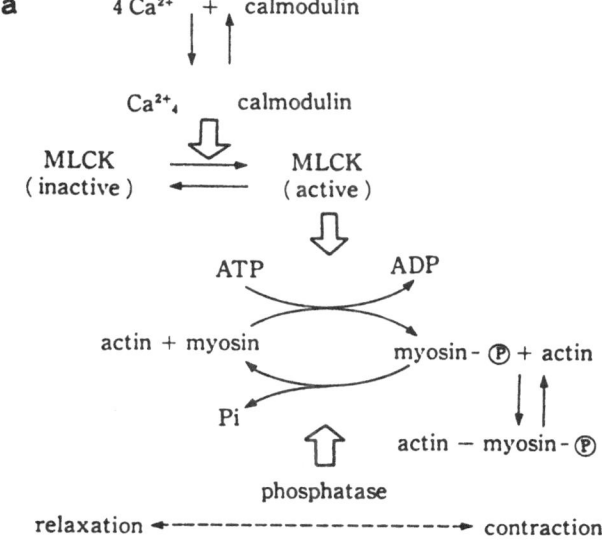

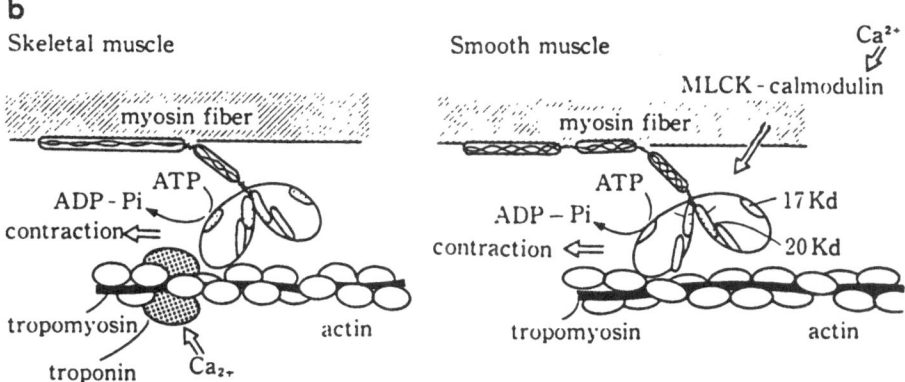

Fig. 1. a Schematic arrangements of myosin phosphorylation theory. **b** Site of phosphorylation of myosin light chain (20 Kd protein of myosin light chain) in smooth muslce and in skeletal muscle. Note: in cardiac muscle, myosin possesses only 6S (narrowing portion), but smooth muscle myosin possesses 10S and 6S sites. Myosin also possesses 17 kD and 20 kD proteins as phosphorylation sites.

structures in these regions is recognized. However, such triads are absent in smooth muscle, and the foot-like structure is found between the sarcolemma and the SR which is located just beneath the sarcolemma. In cardiac and skeletal muscle, a Z-membrane (disc) is also present, whereas in smooth muscle, dense bodies are diffusely present [15].

Other features of smooth muscle cells not seen in skeletal and cardiac muscle are the spindle shape of the cells, the central localization of the nucleus, the cell-to-cell tight junction (nexus), and diffuse innervation with no clear morphological structure at the nerve terminal region. Nerve terminals, including varicosities, are distributed on the adventitial layer in vascular tissues. In smooth muscle cells, the ratio of SR per cell surface is small (about 2%–7% depending on cell type) in comparison with that of skeletal and cardiac muscle (15%–20%) [15]. In cardiac muscle, cholinergic and adrenergic reciprocal innervation is clearly seen as a double innervation. However, in visceral smooth muscle tissues, cholinergic, adrenergic and nonadrenergic-noncholinergic nerves, including excitatory and inhibitory nerves, may all occur [16–20], although vascular tissues, except in a few vascular beds [21], are innervated by only adrenergic nerves [15,17–20,22].

Ionic Channels in Vascular Smooth Muscle

In 1955, Bülbring [23] introduced the microelectrode method for smooth muscle cell research. Since 1980, the voltage- and patch-clamp methods have been successfully applied to dispersed smooth muscle cells or fragmented smooth muscle cell membranes in the same way as they have been applied to other excitable and non-excitable cells [24,25].

K channels. The membrane potential of various vascular smooth muscle cells lies between −45 and −75 mV, and that of guinea pig, rat, rabbit dog and

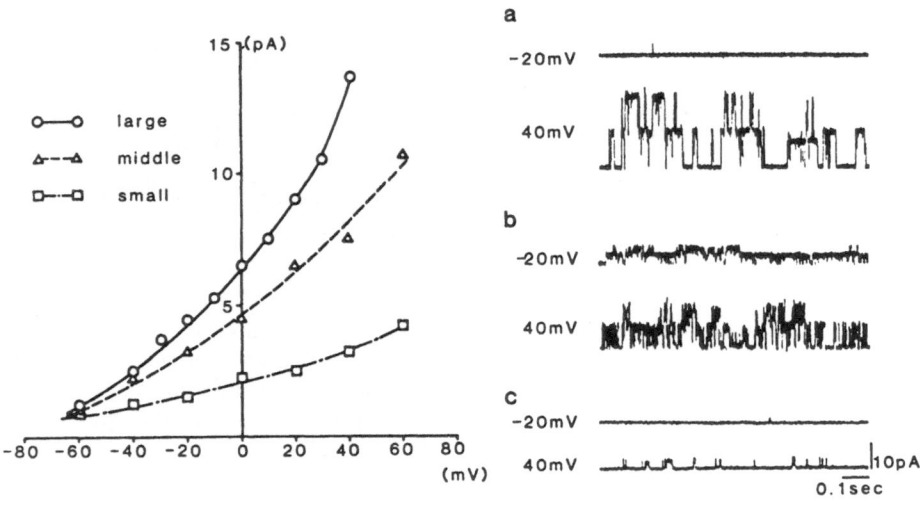

Fig. 2. Three different Ca-dependent K currents, classified according to the amplitude of unitary current and conductances recorded from dispersed smooth muscle cells of the rabbit portal vein. Large (K_L 150 pS) and small (K_S 50 pS) amplitude K currents are sensitive to the cytosolic Ca concentration, but middle amplitude (K_M 100 pS) K current is more sensitive to bath concentration of Ca. Channel conductances of the unitary current were measured under the following conditions; potential range −20 mV to +20 mV, physiological salt solution (PSS) in the pipette 142 mM K, 4 mM EGTA in the bath. inside-out membrane patch. (From [42])

porcine coronary arteries is from -50 to -70 mV. They are electrically quiescent [26–36].

Using the voltage- and patch-clamp procedures, K channels have been classified as either Ca-dependent or Ca-independent K channels in various vascular smooth muscle cells [37–46]. The former is sensitive to intracellular Ca. This intracellular Ca-dependent K channel has been subtyped into K_L and K_S. The former has a large conductance (200–300 pS; measured with high K on both sides of the membrane), and the latter a small conductance (<100 pS). The other Ca-dependent K channel subtype is more sensitive to extracellular Ca than intracellular Ca (K_M channel; the conductance of a unitary current is about 180 pS; [42]). The (K_L and K_M channels are sensitive to tetraethyl-ammonium (TEA), but less sensitive to 4-aminopyridine (4-AP), while the K_S channel is less sensitive to TEA, but is sensitive to 4-AP [38,39,41,42,47,48]. Figure 2 shows three different Ca-dependent K currents (K_L, K_M and K_S) recorded from the rabbit portal vein.

In additon, a Ca-independent K channel of large amplitude has been found in the smooth muscle cells of rabbit ear artery. This channel is insensitive to intracellular Ca, but sensitive to intracellular ATP and glibenclamide (ATP-sensitive K channel; [49]), whereas in the rat portal vein, the K_S channel is sensitive to ATP [50]. Thus, two ATP-sensitive K channels have been recorded from different tissues. In dispersed smooth muscle cells of rat pulmonary artery, a Ca-independent 4-AP-sensitive K channel has been identified using the whole-cell voltage-clamp procedure [45]. However, the nature of the unitary current of the Ca-independent K channel has yet to be studied in detail.

Using the whole cell voltage-clamp procedure, depolarization of smooth muscle cells of guinea-pig ileum or portal vein dispersed with collagenase (holding potential -60mV, command pulse above -40mV) produce a transient Ca influx (I_{Ca}), and a subsequent transient eflux (K_t) and a sustained eflux (K_s). Furthermore, prolonged depolarization provoked oscillatory efluxes (Koo; or spontaneous transient eflux: STOCs) on the K_S [37,43,51–54]. K_t and Koo result from activation of the Ca-dependent K channel, and these currents have a close relation to the K_L classified from the unitary current, but no unitary current corresponding with K_s has yet been found [41,42,52,54] K_t has a causal relation with the generation of a later hyperpolarization of the action potential and with K_L. Koo and K_t are markedly enhanced, but I_{Ca} is inhibited by Ca released from SR [44,51–55].

Ca channels. Ca channels responsible for the generation of Ca influx in vascular smooth muscle have been classified into two subtypes using the whole cell voltage- and patch-clamp procedures [56–59]. One channel possesses a low threshold (more negative potential level), a low amplitude of the unitary current (8–12 pS), a rapidly occurring inactivation, and is less sensitive to dihydropyridine derivatives. This channel is termed the T (transient) type, according to the classification made on dispersed cardiac muscle cells. The second type of Ca channel possesses a high threshold, a large conductance of the unitary current (20–25 pS), slowly occurring inactivation and is sensitive to dihydropyridine derivatives. This channel is termed the L (long-lasting) type. Both the T and L types of Ca channel are voltage-dependently modified. In addition, in ganglion cells and neurons, but not in cardiac and smooth muscle cells, another type of Ca channel has been identified form conductance measurement (N channel: neither L nor T type; [60]. This channel has an intermediate conductance (12–15 pS), its rate of inactivation is moderate, and it is less sensitive to dihydropyridine derivatives.

In many vascular and visceral tissues, two different Ca channels (L and T types) have been identified from the amplitude and conductance of the unitary currents, and also from their sensitivity to dihydropyridine derivatives (aorta [61], mesenteric artery [62], ear artery [63], saphenous vein [64], taenia coli [65]). Figure 3 shows two voltage-dependent Ca currents recorded from the basilar artery. As described previously, there is spontaneous electrical activity in the portal vein, but in many resistance vessels the action potential has to be evoked by a depolarization (the excitatory junction potential EJP [66–69]), which can be induced by various procedures, such as perivascular sympathetic

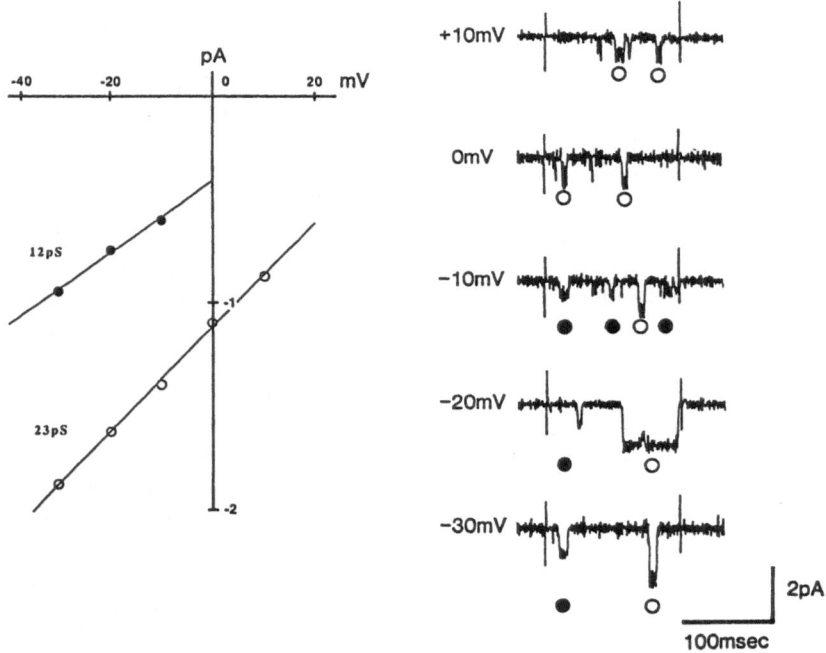

Fig. 3. Two different amplitudes of voltage-dependent Ca current (L and T types) recorded from dispersed smooth muscle cells of the rabbit basilar artery, using the cell-attached patch clamp procedure. *Left*: Current-voltage relationships observed from two different amplitudes of Ca currents; large and small amplitudes of I_{Ca} showed slope conductances of 23 and 12 pS, respectively. *Right*, actual traces of inward currents recorded at different membrane potentials (10,0, −10, −20, −30 mV); *open circles*, openings of the 12 pS Ca channel; *closed circles*, openings of the 23 pS Ca channel (Y Inoue, M. Oike, K. Kitamura, unpublished observations)

nerve stimulation, direct electrical stimulation [26,27], or by application of agonists [35,48,70,71]. This action potential is generated by activation of the voltage-dependent Ca channel. The main action current is carried through the L type of Ca channel, but the role of the T type has not yet been clarified [57–59,61–65,72]. The L type of Ca channel distributed in smooth muscle cell membranes may not have exactly the same features as those found in cardiac muscle cells, because cyclic AMP perfused into the smooth muscle cell or β-adrenoceptor stimulation did not enhance the channel activity [43], and the site of action of D600 differed in the two tissues [52,55]. Recent investigations in visceral smooth muscle cells including those of vascular tissue, have classified the voltage-dependent Ca channel into large, middle, and small conductance Ca channels ($I_{Ca}L$, $I_{Ca}M$, and $I_{Ca}S$, respectively). $I_{Ca}L$, but not $I_{Ca}S$ was consistently inhibited by dihydropyridine derivatives, and $I_{Ca}M$ of some vascular smooth muscle cells were inhibited, and of other cells were not inhibited by these agents [73].

Ca influx is triggered by stimulation of various receptors distributed on the sarcolemma. Such a channel is called a receptor-activated (operated) Ca channel [11,74]. In the rabbit ear artery, Benham & Tsien [75] reported activation of the purinergic receptor by nonselective ATP-induced channel opening for Na and Ca (assumed to be in the ratio of 3:1, and in physiological condition, 10:1). Presumably, this receptor-operated channel was a nonselective cation channel. Activation of the receptor may directly or indirectly modify the voltage-dependent Ca channel [45,76,77]. It is certain these modifications of the Ca channel by agonists require GTP-binding proteins.

Na channels. In the rat pulmonary artery and portal vein, when the holding potential was set at -80 mV, but not when it was at -60 mV, depolarization of the membrane (above -20 mV)provoked a transient influx (5–10 msec) measured with the whole cell voltage-clamp procedure. This current was generated in Ca-free solution containing 1 mM EGTA, or a few mM Mn. When the concentration of Na in the bath was reduced, or a very low concentration of tetrodotoxin was added to the bath, this influx ceased (the IC_{50} value for tetrodotoxin is 8 nM). Therefore, this influx was thought to be a Na current (I_{Na}; [78]). I_{Na} was enhanced by application of chloramine-T, an inhibitor of Na-channel inactivation, and was not inhibited by dihydropyridine derivatives. When the current-voltage relationship was observed in the absence of Ca in the bath, the peak amplitude of the influx shifted to the left compared to that of 2.5 mM Ca [78]. The physiological role of this Na current is not yet known since the membrane potential of these vascular smooth muscle cells was approximately -50 to -60 mV, and 90% of this Na current was readily inactivated at -60 mV. Figure 4 shows the Na current (I_{fast}) and the effects of tetrodotoxin on the Na current. In some visceral smooth muscle cells, the presence of another type of Na channel has been reported. Amedee et al. [79] reported the presence of a less tetrodotoxin – sensitive Na current (IC_{50} 2 μM) in rat myometrium; this Na current had similar properties to those recorded from the rat azygos vein [80].

In addition, the Na current can be recorded through the Ca channel. In the rabbit intestine, a depolarization pulse can evoke an action potential with a plateau in nominally Ca-free solution. However, this current proved

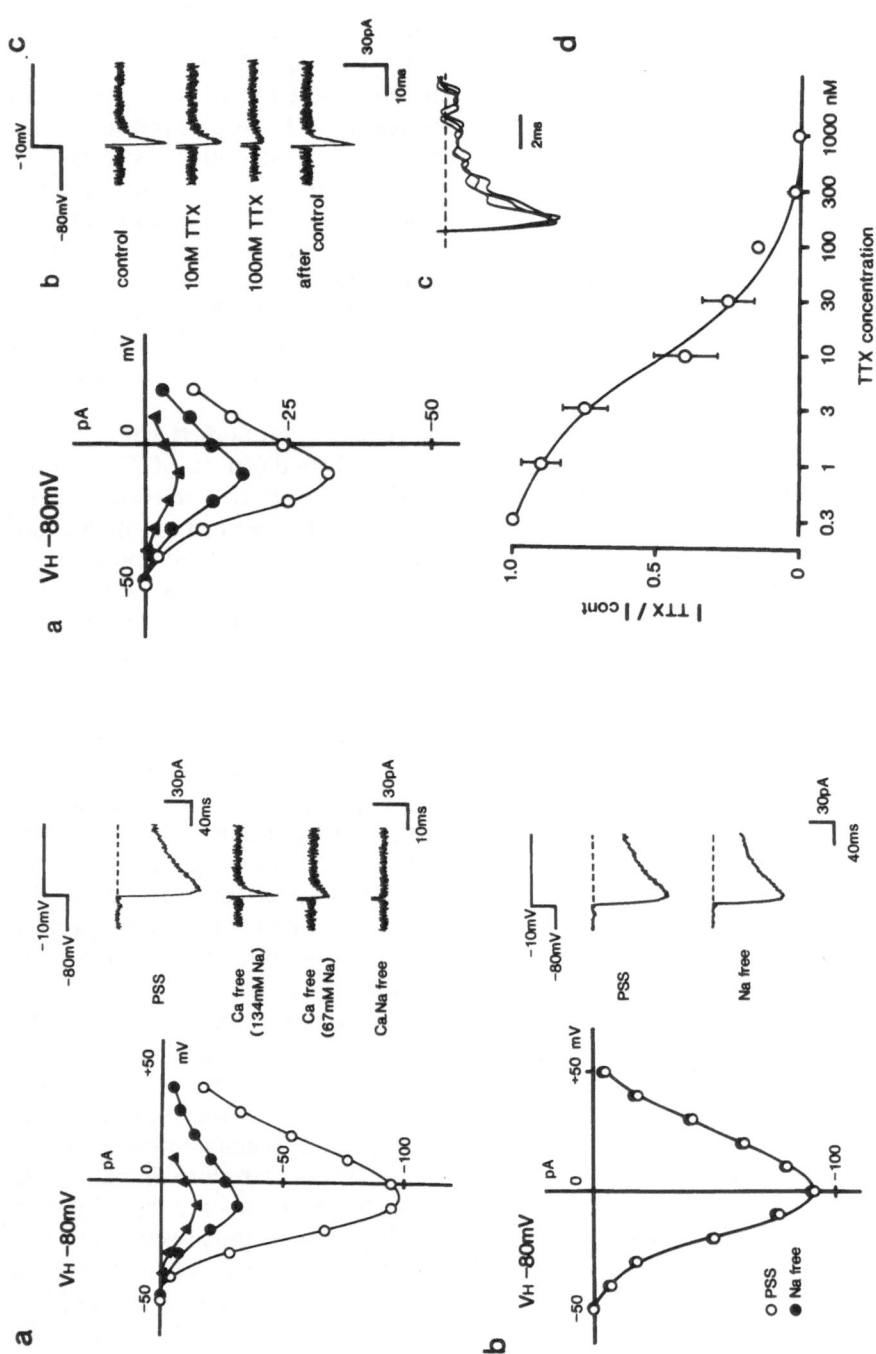

Fig. 4. Tetrodotoxin-sensitive Na current recorded from smooth muscle cells of the pulmonary artery. Current-voltage relationships of a fast inward current (I_{fast}), and b slow inward current (I_{slow}). c, d Effects of tetrodotoxin on I_{fast} recorded in a Ca-free solution containing 2.5 mM Mn, and 134 mM Na. A The membrane potential was maintained at −90 mV. The peak amplitude of influx recorded in PSS, Ca-free solutions with 2.5 mM Mn, 134 mM Na and 2.5 mM Mn, 67 mM Na, were plotted; b The peak amplitude of the influx recorded in PSS and Na-free solution were plotted; c The current-voltage relationships of I_{fast} recorded in the presence or absence of tetrodotoxin; d The relationship between tetrodotoxine concentrations and the relative amplitude of I_{fast} evoked by a command potential to −10 mV from the holding potential of −80 mV. (From [78])

insensitive to tetrodotoxin (5 μM), but was blocked by dihydropyridine derivatives [44]. This means that the ion selectivity of this Ca channel is lost in the presence of low concentrations of Ca (below 0.1 mM).

Factors Modifying Ionic Currents and Mechanical Responses in Vascular Smooth Muscle Cells

Neural Factors

Vascular smooth muscles are innervated mainly by adrenergic fibres. However, Brayden and Large [47] reported possible vasodilator actions of cholinergic fibres on the rabbit lingual artery, and vasodilator actions induced by cholinergic fibres have also been reported in the uterine artery [81,82]. From in vivo experiments, Nakazato et al. [83] also reported that muscarinic fibres contributed to vasodilation. Therefore, cholinergic innervation cannot not be excluded.

In many resistance vessels, perivascular adrenergic nerve stimulation can evoke an EJP with or without the generation of a slow depolarization (e.g. guinea pig mesenteric artery [69,84]. Figure 5 summarizes the electrical responses evoked by perivascular nerve stimulation in various vascular tissues. EJP and slow depolarization do not occur following application of tetrodotoxin. When repetitive perivascular nerve stimulation at high frequencies (above 1 Hz) is applied, there is summation of EJP, and when depolarization exceeds the threshold, action potentials (spike potentials) occur on the EJPs [21,29,33, 66–69,85–93]. With application of dihydropyridine derivatives, the spike potential is blocked, but EJPs persist [36,69]. EJP is not modified by application of α_1-adrenoceptor blockers, but the slow depolarization can be blocked by α_1-and/or α_2-adrenoceptor blockers [85,87,88]. The EJP amplitude was markedly modified by purinergic II-receptor blockers or by desensitizing agents [70,94–102]. Following treatment with reserpine, EJP gerneration still occured, in spite of a decrease in the amount of noradrenaline (NA) to 1% of the control value, but the slow depolarizartion was blocked [93]. On the other hand, guanethidine blocked the generation of EJPs [93,100]. When ATP was applied using an ionophoretic procedure, a rapid depolarization occurred with a shape resembling that of the EJP [100,102–104]. Thus, it seems likely that

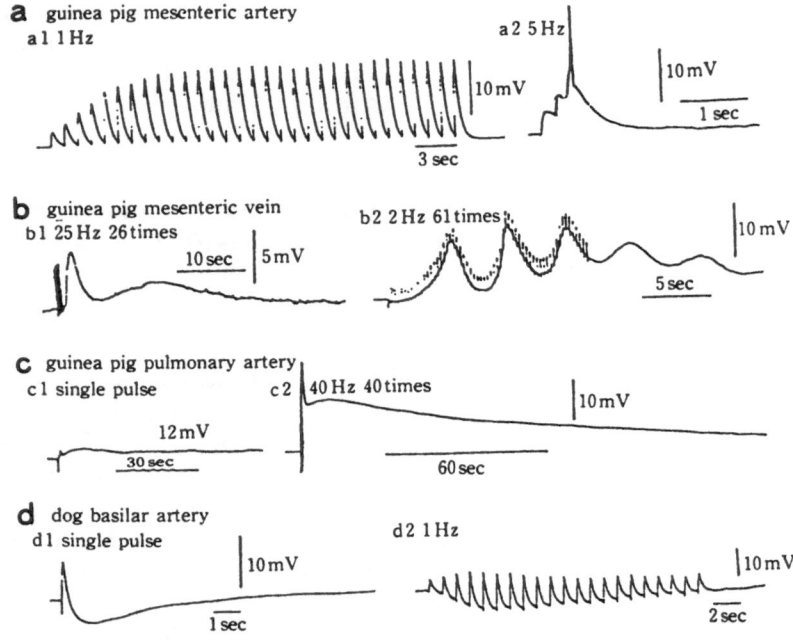

Fig. 5. Electrical responses evoked by perivascular nerve stimulation in various vascular beds. Different frequencies of perivascular stimulation were applied to various vascular tissues. (From [36])

the EJP is generated via activation of the PII receptor by ATP, a co-transmitter of NA, and that the slow depolarization is due to activation of α_1-or α_2-adrenoceptor [19,20,22]. However, there is still dispute as to whether the EJP. is generated by activation of γ-adrenoceptor(adrenoceptor resistant to classical blockers) [87,88] or PII (ATP) receptor [18–20,22,104]. In the facial vein, perivascular nerve stimulation is reported to activate β-adrenoceptor and cause hyperpolarization of the membrane. This hyperpolarization can be blocked by propranolol but not by α_1- or α_2-adrenoceptor blockers (prazosin, phentolamine, or yohimbine). However, in this tissue, the presence of α-adrenoceptor has been reported, i.e. after application of propranolol, NA depolarized the membrane by acting on α_1-adrenoceptor [48].

Prejunctional nerve terminals contain many receptors and these receptors regulate release of the transmitter, in a positive or negative manner. Thus, α_2-adrenergic, H_3-histaminergic, PGI_2-, and M_2-muscarinic receptors have negative effects, whereas β-adrenoceptor positively regulates release of neurotransmitters, as estimated by either NA outflow [92,105–109], or the amplitude of the EJPs [31,32,84–86,91,93,102,104,108,110–113]. Regulation of transmitter release from prejunctional nerve terminals (varicosities) has been confirmed in bioassay and electrophysiological investigations, [108]. In some vascular beds, the presence of α_1-adrenoceptor on nerve terminals has been reported [107,108].

In many elastic arteries and veins, neither repetitive perivascular nerve stimulation nor exogenously applied NA depolarizes the membrane, but they still produce contraction. Such responses are called "pharmaco-mechanical coupling" [36,90,114–119]. Now, it is clear that in visceral smooth muscle, inositol 1,4,5-trisphosphate (IP_3) which is synthesized by activation of α_1-adrenoceptors, releases Ca from the SR [119–125]. Thus, NA, a neurotransmitter, may increase Ca in the cytosol and produce contraction [126–134]. Generation of the action potential, EJP, and slow depolarization induced by NA and/or ATP in vascular smooth muscle may produce cumulative actions which increase cytosolic Ca.

Endothelium-Derived Contracting and Relaxing Factors

Endothelial cells release various substances and thus regulate vascular smooth muscle tone. In 1980, Furchgott and Zawadzki [135] found a factor that relaxed contracted vascular tissues and this factor was termed "endothelium-derived relaxing factor" (EDRF) [135–138].Relaxation induced by EDRF is due to synthesis of cyclic GMP in the cytosol, in the same manner as synthesis of cyclic GMP occurs on application of nitro-compounds, such as nitroglycerine, isosorbide dinitrate or nitroprusside. Synthesis of cyclic GMP induced by EDRF and nitro-compounds is inhibited by methylene blue or hemoglobin [139–147]. Further evidence has accumulated to clarify the nature of EDRF and at present, nitric oxide (NO) is thought to be a candidate for EDRF [99,139,142,148,149]. Recently, it was suggested that a source of NO which could stimulate synthesis of cyclic GMP might be L-arginine, but not D-arginine [148–150]. This action of L-arginine on the synthesis of NO and the relaxation of contracted tissues can be inhibited by L-G-monoamino-methyl-arginine [149,151].

When ACh is applied to intact contracted vascular tissues (endothelium preserved), they relaxed and this relaxation is mediated by release of EDRF following stimulation of muscarinic 2 (M_2) receptor distributed on the endothelial cell. By contrast, stimulation of M_1 receptor distributed on the endothelial cell releases an endothelium-derived hyperpolarizing factor (EDHF), which hyperpolarizes the smooth muscle membrane [32,33,99,152–154]. EDHF is insensitive to methylene blue and hemoglobin, and is not related to synthesis of cyclic GMP. Standen et al. [49] postulated that hyperpolarization induced by EDHF or K channel openers (cromakalim or pinacidil) is due to activation of the Ca-independent ATP-sensitive K channel. This channel differs from the channel responsible for the nicorandil-induced hyperpolarization reported in rat portal vein because the latter channel is sensitive to Ca and shows a much smaller conductance of the unitary current [50]. EDHF has also been found to be released by histamine [154]. Schematic arrangements of endothelial and smooth muscle cells, in relation to EDRF, EDHF, PGI_2, and endothelin are shown in Fig. 6.

PGI_2 is mainly synthesized in endothelial cells [100, 155]. This substance is released from endothelial cells, synthesizes cyclic AMP in smooth muscle cells,

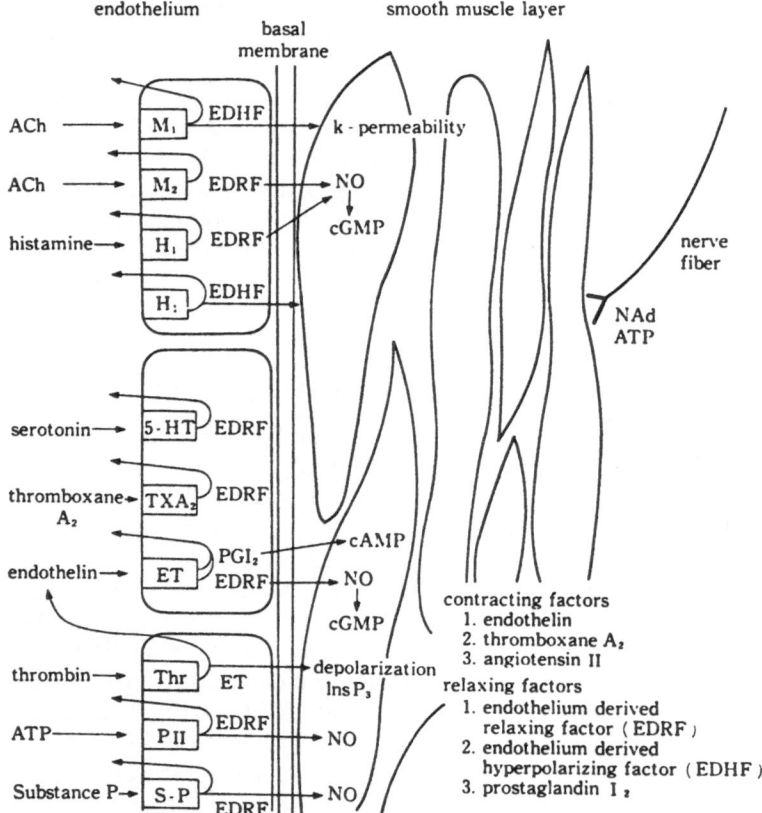

Fig. 6. Schematic functional arrangements of endothelial cells and smooth muscle cells in relation to release of EDRF, EDHF, PGI$_2$ and endothelin by various stimulants. NO is a product of EDRF (presumably L-arginine)

and produces relaxation of resting and contracted vascular tissues, in the same manner, following stimulation of β-adrenoceptor.

In contrast to these three vasodilating factors (EDRF, EDHF, and PGI$_2$), the presence of vasoconstrictor factors such as endothelin [156] and thromboxane A$_2$-like substance [157] has been reported. Yanagisawa et al. [156] first described endothelin (now 3 subtypes; I, II, and III derivatives), a substance which can induce prolonged vasoconstriction of vascular beds. However, in the presence of endothelial cells, endothelin first produced relaxation, and then contraction of vascular tissues, whereas in endothelium-denuded tissues a larger contraction occurred. Relaxation induced by endothelin is thought to be due to release of EDRF from endothelial cells and this relaxation ceases following application of oxy-haemoglobin [62,98,153,158–160].

Vasoconstriction induced by endothelin is thought to be due to increased activation of voltage-dependent Ca channels [161], and also to an increase in the synthesis of IP$_3$ [158,162]. Increased cytosolic IP$_3$ may cause contraction ,but increased Ca may accelerate inactivation of the voltage-dependent Ca channel following release of Ca from the SR. It has been reported that endothelin activates both L and T channels [73].

Actions of Ca-Regulating Substances in Relation to the Roles of SR

The SR plays an essential role in the regulation of the concentration of free Ca in the cytosol of vascular smooth muscle cells. In the guinea-pig mesenteric artery, action potentials evoked by electrical stimulation generated contraction. Following application of procaine, action potentials were still evoked, but they did not produce contraction, i.e. procaine decouples excitation-contraction coupling. Procaine is known to inhibit the K permeability of the membrane, and also to release Ca from SR by blocking the Ca-induced Ca release mechanism in vessels [90,117,163–167]. In smooth muscle cells skinned with saponin, caffeine and IP$_3$ could release Ca from th SR and produce contraction [7,165], and this Ca-releasing action can be inhibited markedly by procaine in the porcine coronary artery [117]. Furthermore, heparin inhibited the release of Ca induced by IP$_3$ [168], and ryanodine inhibited both Ca-induced, and caffeine-induced release of Ca [169,170]. In fact, caffeine and ryanodine transiently acceleated then blocked the generation of the unitary K current (K$_L$) and Koo [54], as estimated from the voltage and patch-clamp procedures. Thus, at least two different Ca-releasing mechanisms may occur in the SR. The amino acid sequences of the ryanodine-binding [171] and IP$_3$-binding proteins [172] in SR have been determined. In addition, GTP-sensitive Ca release in smooth muscle cells has been reported [173,174].

SR has the ability to accumulate Ca through activation of Ca-ATPase in the presence of phospholambane, ATP, cyclic AMP, and other substances. Therefore, phosphorylation of the Ca channel in the SR plays some role in the active uptake of Ca. The Ca-ATPase distributed in SR differs from that in the sarcolemma [151,175–178].

Contraction of vascular smooth muscle is thought to be due to an increase in free Ca above 0.1 μM, with the maximal amplitude of contraction being evoked by 5–10 μM Ca [117]. We postulate that Ca entering the cell through the voltage-dependent Ca channel may be directly incorporated into the SR, and the influx of Ca may activate the Ca-induced Ca release mechanism (caffeine and ryanodine-sensitive mechanism). However, IP$_3$ releases Ca from the SR in a different manner (heparin-sensitive mechanism) [173]. Even the sites of release of Ca from the SR differ; it is not yet clear whether the sources of Ca utilized for Ca-induced and IP$_3$-induced release are the same, and further studies are needed.

Second messengers, such as cyclic nucleotides, IP$_3$ and diacylglycerol (DG) modulate cytosolic Ca either by accumulation of Ca into the SR, or by Ca extrusion from the sarcolemma by activation of the Ca-ATPase at the SR or sarcolemma. Cyclic nucleotides are therefore negative controllers, and IP$_3$ is a positive controller of the amount of free Ca. However, there is still uncertainty as to the role of protein kinase C activated by DG, a co-product with IP_3 in the presence of Ca with phosphatidic acid in smooth muscle cells and nerve terminals. Phorbol esters (tumor-promoting substances) reduce the amount of IP_3. As a consequence, the amount of released Ca in the cytosol is reduced [167,179]. However, low concentrations of phorbol esters have been found to enhance contraction with a slight increase in the cycling rate of actin-myosin cross bridge formation [179]. Presumably, an increased Ca sensitivity of the contractile protein may partly contribute to enhancement of phorbol esters. Phorbol esters can also inhibit or accelerate the channel activity of I_{Ca} or I_{Na} in many types of cells [131–133]. Furthermore, activation of protein kinase C has been reported to contribute to the generation of a latch phenomenon (contraction sustained under conditions of low Ca, low phosphorylation of myosin light chain, and a low cycling rate of cross bridges between actin and myosin); [180–183] through interaction with caldesmon.

Specific Features of Coronary Artery

Figure 7 demonstrates regional differences in mechanical responses recorded from rabbit coronary artery [184]. Tissues were excised from right (RCA), left anterior descending (LAD), and left circumflex (LCX) coronary arteries, and responses to 128 mM K, 10 μM acetylcholine (ACh), and 30μM 9,11-epithio-11, 12-n-methano-thromboxane A_2 (a stable derivative of thromboxane A_2, STA_2) are show. To compare mechanical responses, fine strips of the same size were prepared (length 0.15–0.2 mm, width 0.05–0.06 mm, thickness 0.04–0.05 mm). In the presence of endothelial cells, RCA and LCX produced much larger

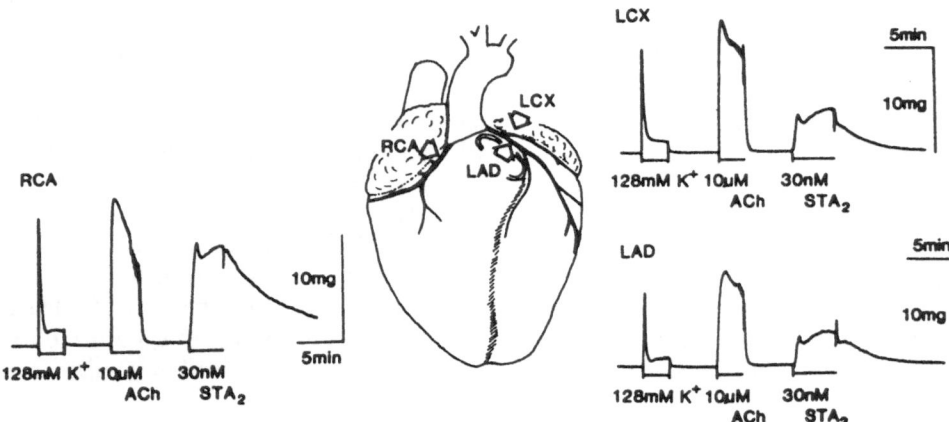

Fig. 7. Surface branches of coronary artery in the guinea pig heart, and regional differences in response to stimulants. RCA, right anterior descending coronary artery; LAD, left anterior descending coronary artery; LCX, left circumflex coronary artery (From [184])

K^+-induced contractions than LAD, whereas STA_2 produced larger contraction of LCX or LAD than of RCA. ACh produced much larger contraction than K^+ in all three regions. After the endothelial cells were removed, the mechanical responses to both ACh and STA_2 were markedly enhanced, whereas high $[K^+]$ only enhanced the tonic response, but not the phasic responses [184]. Regional differences have also been investigated in guinea pig coronary artery [154]. When mechanical responses evoked by high $[K^+]$ (118 mM), ACh (10 mM), and STA_2 (10 nM) were compared for the proximal (0.2 mm outer diameter) and distal (0.01 mm outer diameter) regions of LAD, it was found that ACh and STA_2 produce much smaller contractions in the proximal region than did K^+ in the presence of endothelial cells. However, in the distal region, the ACh-induced contraction was much smaller than the K-induced contraction, but STA_2 generated a similar amplitude contraction to that observed on application of high $[K^+]$. Following the removal of endothelial cells, STA_2-induced contraction was larger than that evoked by high $[K^+]$ in both regions. By contrast, after removal of endothelial cells, tissues excised from the distal region produced much larger contractions to ACh than to K^+ approximately twice, but in the proximal region, ACh produced a contraction only half the amplitude of that induced by K^+ [154].

Species differences in the muscarinic receptor distribution in coronary arteries have been inferred from the membrane potential change induced by application of ACh. In porcine and guinea-pig coronary arteries (endothelial cells attached), ACh (10 mM) hyperpolarized the membrane (due to release of EDHF), whereas following removal of endothelial cells, ACh depolarized the membrane in guinea-pig coronary artery, but in porcine coronary artery provoked no electrical events [27,165,185]. However, in the coronary arteries of rabbit and dog, ACh consistently depolarized the membrane whether in the presence or absence of endothelial cells [104,186]. In the procine main coronary artery, NA only activated β-adrenoceptor, but not α-adrenoceptor [27].

Drug Actions on Coronary Artery Vasospasm

Reversible vasospasm in the coronary artery, associated with pain (angina pectoris) may occur in patient during rest. The pathogenesis of this disease has been investigated intensely. Pathophysiological changes in endothelial cells with increased sensitivity of receptors (in particular, the histaminergic receptor) distributed on muscle membranes have been postulated to be factors which induce reversible vasospasm [157]. Endothelial cells regulate muscle tone, as described previously. Vasoconstrictor substance, such as thromboxane A_2 or thromboxane A_2-like substance released from platelets, endothelial cells, or thrombosis in sclerotic vessels may produce contraction in smooth muscle. A reduction in the synthesis of relaxing substances, such as PGI_2, EDRF, and EDHF from regenerated endothelial cells after they have suffered damage, or from aged endothelial cells may also induce an increase in muscle tone. It may be that during coronary vasospasm vasocontstrictor factors are released preferentially in comparison with relaxing factors. Neurotransmitters released from varicosities of sympathetic nerves mainly regulate muscle tone from the adventitial site and the deep layer of the tunica media so that endothelial cells may not be directly regulated by such neural factors (NA and ATP) under physiological conditions. A reduction in the density or sensitivity of the α_2-adrenoceptor on nerve terminals, and increased sensitivity and number of α_1- and/or α_2-adrenoceptors on the sarcolemma may trigger vasospasm. However, pharmacological experiments demonstrated the absence of the α-adrenoceptor in large branches of the coronary artery in some species [28]. Changes in the Ca sensitivity of contractile proteins (calmodulin) and other factors (20 Kd proteins, actin-myosin interactions, myosin light chain phosphatase, etc.) in coronary muscles themselves may also contribute to the generation of vasospasm.

Nitro-(Including Nitroso- and Nitrite-) Compounds

Nitro-compounds such as nitroglycerin, amylnitrate, isosorbide dinitrate, and nitroprusside are well documented drugs for the alleviation of angina pectoris. These agents contain a NO_2 residue, and this NO_2 is transformed to nitric oxide (NO) in the cytosol, and synthesizes cyclic GMP [139–146,169]. Such a pathway for the synthesis of cyclic GMP also occurs with EDRF [100,151,187]. Hence. EDRF is thought to be an endogenous nitro-compound, as described previously. Synthesis of cyclic GMP occurs on application of atrial natriuretic peptide to the sarcolemma [145,151,187]. Methylene blue or hemoglobin inhibits the synthesis of cyclic GMP in the cytosol induced by nitro-compounds and EDRF but not in the sarcolemma induced by atrial natriuretic peptide [139,149]. Thus, the synthetic pathway of cyclic GMP induced by atrial natriuretic peptide differs from that induced by nitro-compounds or EDRF. Cyclic GMP synthesized through both paths reduces the concentration of free Ca in the cytosol [168,188], due to an increase in Ca extrusion through activation of Ca-ATPase at the sarcolemma [189,190]. Vrolix et al. [178] and Raeymaekers et al. [15] have reported that cyclic GMP accelerated the Ca pump in an indirect manner through phosphorylation of phosphatidyl-inositol. Eggermont et al. [175,176] reported that in the bovine pulmonary artery, the Ca pump (via activation of Ca ATPase, 130 Kd proteins) in the sarcolemma may play a more minor role than Ca uptake in the SR. However, Rüegg [191] reported that cyclic GMP, as well cyclic AMP, phosphorylated myosin light chain kinase. As a consequence, Ca-calmodulin complex-induced phosphorylation of 20 Kd proteins of myosin light chain was inhibited, causing relaxation. These observations are supported by Yanagisawa et al. [192]. They found that nitroglycerin does not modify the cytosolic Ca measured using fura-2, but relaxes the precontracted canine coronary artery by excess K, but not by agonists. However, these observations on the phosphorylation of myosin in physiological conditions are not completely supported by other investigators [11,191,193]. Nitroglycerin reduces marginally blood pressure, however, it causes dilatation of coronary arteries, and reduces venous return to the heart. In addition, it increases the anastomoses between coronary vascular beds, thus reducing the cardiac load. Such concomitant actions of nitro-compounds may be important during anginal attacks.

β-Adrenoceptor Blockers

β-Adrenoceptor blockers are commonly used for the prevention of angina pectoris. In many coronary arteries, activation of β-adrenoceptors in coronary vascular smooth muscle cells produces hyperpolarization of membranes and relaxation of tissues due to synthesis of cyclic AMP [194]. Therefore, β-adrenoceptor blockers may not act directly as vasodilators. Presumably, these agents may inhibit cyclic AMP-induced augmentation of I_{Ca} in cardiac muscle cells, and also inhibit the release of renin and prevent vasomotor actions via the medulla oblongata. In addition, inhibition of β-adrenoceptors present on pre-junctional nerve terminals may reduce the release of excitatory transmitters [69,113,195]. Thus, the prevention of angina pectoris by β-adrenoceptor blockers may not be due to direct actions on vascular smooth muscle tissues, including that of the coronary arteries.

The role of the β-adrenoceptor in vascular beds is not yet properly understood. However, in tracheal muscles, isoprenaline has been found to produce relaxation through synthesis of cyclic AMP. This second messenger reduced the free Ca in the cytosol, and also hyperpolarized the membrane by phosphorylation of the Ca-dependent K channel [196].

Ca Antagonists

Ca antagonists are often used for the prevention of angina pectoris. These drugs are known to inhibit voltage-dependent L-type Ca channels. To activate this channel, depolarization of the membrane is required. In most vascular tissues, the resting membrane potential is reported to be $-60--70\,mV$. Thus, some L type Ca channels are inactivated, and most T type Ca channels are inactivated at resting membrane potential. Dihydropyridine derivatives such as nifedipine, nisoldipine, or nicardipine, and also other groups of antagonists such as diltiazem and flunarizine, act more potently in the inactivated state than in the resting state of the L type Ca channel, but verapamil acts more on the open Ca channel, as an open channel blocker [65,98,123,197–202]. Therefore, even at the resting membrane potential level, Ca antagonists may have some action on Ca channels (in particular, dihydropyridine derivatives), but when the membrane is depolarized by pathological situations (open or inactivated state of the channel), they may have a more potent inhibitory action on the influx of Ca. Therefore, the sensitivity of Ca channels to Ca antagonists differs according to the state of the channel. Inhibition of the L type Ca channel in vascular smooth muscle cells occurs by a lower probability of the channel being open, with increase in the blank sweeps with no change in the amplitude of the unitary current amplitude. In addition, Ca antagonists act not only on vascular tissue, but also on cardiac muscle tissue, and prevent excess demand for oxygen. Figure 8 shows the effects of nifedipine on the L type Ca channel recorded using patch

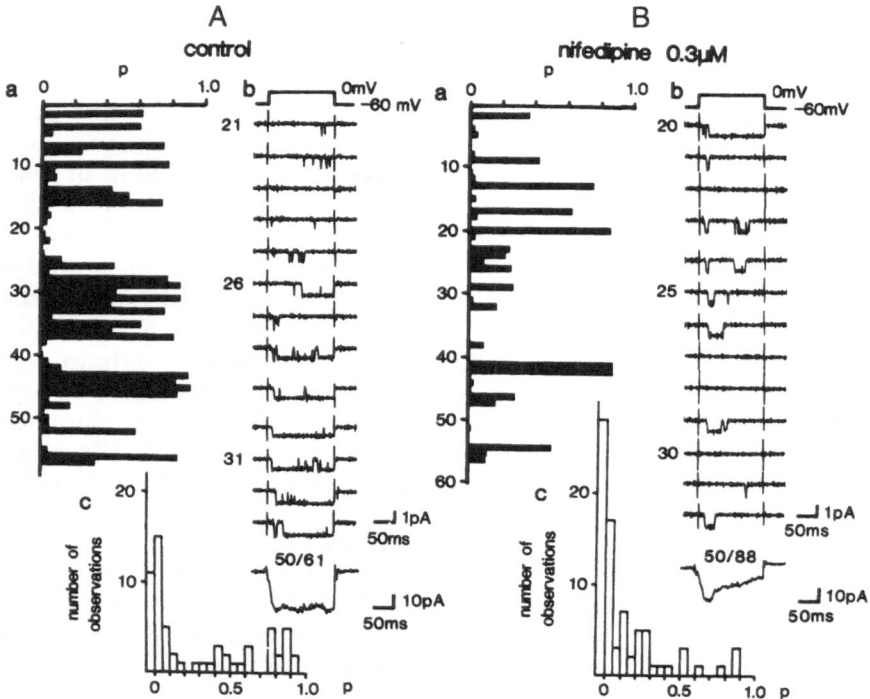

Fig. 8. Effects of nifedipine on the unitary Ca channel current (L type) recorded from dispersed smooth muscle cells of the rabbit portal vein using the patch clamp procedure. Electrical responses were evoked by two different frequencies of stimulation. Recording conditions: both pipette and bath solutions contained high [K], holding potential was $-80\,mV$, depolarization potential and duration were $0\,mV$ and $150\,ms$, respectively; stimulus frequency was $0.033\,Hz$. **A** control; **B** in the presence of $0.3\,\mu M$ nifedipine; **a** changes in the probability of channel opening in sequential current sweeps; **b** samples of sequential sweeps of the unitary Ba current. Bottom traces show the summated unitary Ba currents of 61 (control) and 88 (nifedipine) sweeps. The fraction of 50/61 and 50/88 indicate the number of sweeps in which unitary Ba currents appeared (50 sweeps out of total proportion of 61 and 81 sweeps). From [73])

clamp procedure. However, a dihydropyridine derivative, Bay K 8644, possesses an agonistic action on the voltage-dependent Ca channel in vascular smooth muscle cells, i.e. Bay K 8644 enhances the probability of the channel being open, prolongs the opening time of the L type Ca channel, and reduces the blank sweep with no change in the unitary current amplitude (Fig. 9). However, it should be mentioned here that many dihydropyridine-derivative Ca-antagonists possess agonistic actions, and those agonists possess antagonistic actions.

K Channel Openers (Activators)

Nicorandil (2-nicotinamido ethylnitrate) contains NO_2 and acts as a nitro-compound. In addition, this agent acts as a K channel opener, similar to

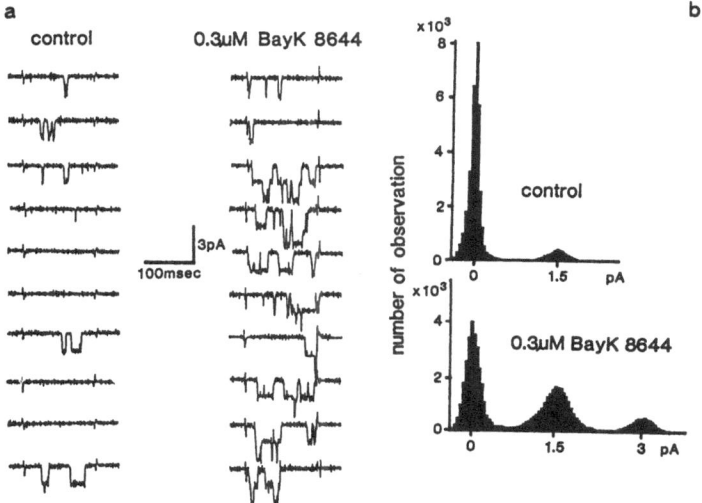

Fig. 9. Effects of Bay K 8644 (0.3 µM) on the voltage-dependent Ca channel (L type) in dispersed smooth muscle cells membrane using the cell-attached patch-clamp procedure. **a** actual traces of the action of nifedipine on the L type Ca channel (holding potential −60 mV, depolarization pulse 130 ms and 0 mV, pipette solution 2.5 mM Ca with TEA and Cs, bath solution 118 mM high [K]). **b** Amplitude histograms of the unitary current of the L type Ca channel. *Upper* before application, *lower* after application of Bay K 8644 (Y. Inoue and K. Kitamura, unpublished observations)

cromakalim and pinacidil [203–205]. Nicorandil hyperpolarizes smooth muscle cell membranes of coronary arteries; with high (more negative) membrane potentials hyperpolarization is less, and with low (more positive) membrane potentials hyperpolarization is greater. Thus the amplitude of hyperpolarization depends on the membrane potential in relation to the K-equilibrium potential. When the membrane is depolarized with high [K$^+$], it is not longer hyperpolarized by nicorandil, but the tissues are relaxed by a nitroglycerine-like action [28,103,172,188,206,207]. Nicorandil-induced hyperpolarization has been reported to block the action potential generated by electrical stimulation in the presence of tetraethylammonium in porcine coronary arteries [206]. This nicorandil-induced hyperpolarization is thought to be due to activation of K_S (Ca-dependent and ATP-sensitive K channel with small conductance). Figure 10 shows the effects of nicorandil on membrane potential and unitary K channel currents. This nicorandil-sensitive K_S channel was sensitive to intracellular Ca and ATP [50]. In addition, this K_S channel activity was blocked by glibenclamide, an anti-diabetic agent and a selective inhibitor of the ATP-sensitive K channel. Charybdotoxin (extracted from *Leiurus quinquestriatus hebraeus* scorpion venom: a blocker of the Ca-dependent K channel with large conductance) did not modify K_S, but markedly inhibited the K_L channel. Thus the action of nicorandil in rat and rabbit portal veins was not prevented by the presence of charybdotoxin (S. Kajioka and K. Kitamura, personal communications). Pinacidil had much the same actions as observed by applications of

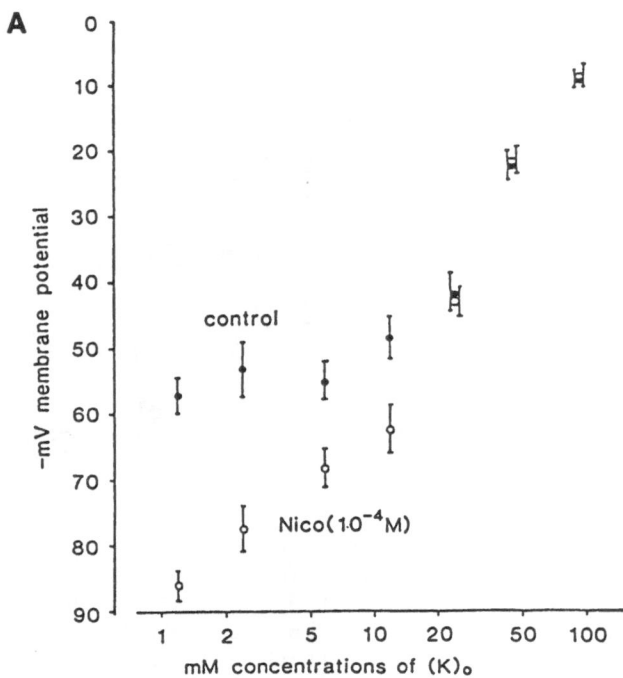

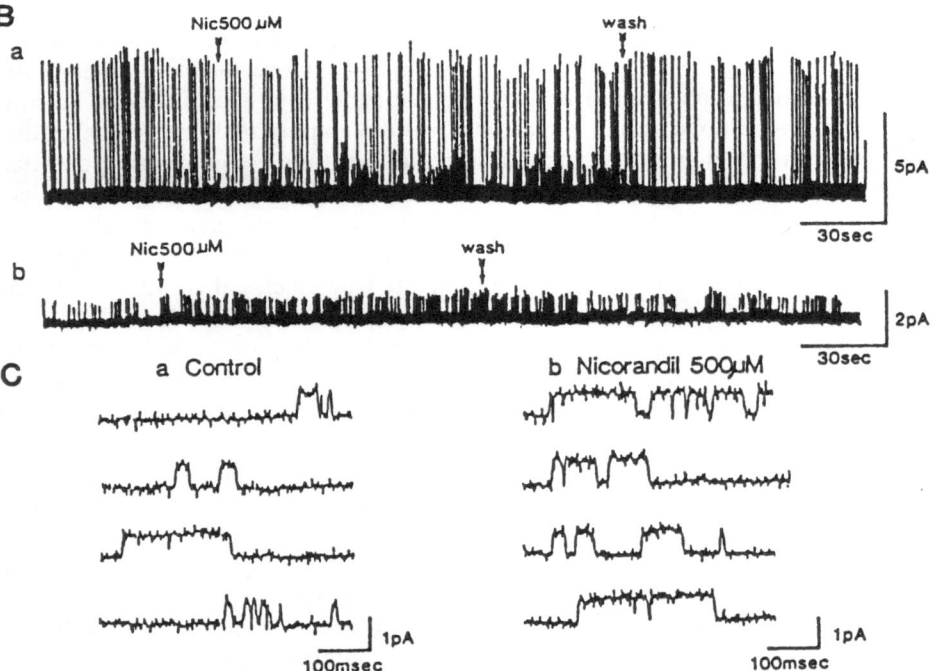

nicorandil on the rabbit portal vein (S. Kajioka and K. Kitamura, personal communications). On the other hand, Standen et al. [49] observed the effects of cromakalim on the unitary K current, and postulated that cromakalim activates the Ca-insensitive ATP-sensitive large amplitude K current. Gelband et al. [208] reported that in rabbit aorta, cromakalim activates the Ca-sensitive large conductance K channels (K_L) within the planar lipid bilayers. Thus, the features of the ATP-sensitive and K-channel opener-sensitive channel activated by nicorandil and cromakalim in vascular beds may show both regional and species variations.

Conclusion

In this article, we have briefly surveyed recent investigations on membrane properties (ion channel properties) of vascular tissues, in relation to the actions of adrenergic innervation and of endothelial cells. In addition, the clinical actions of some drugs on coronary vasospasm were reviewed. Muscle tone, including contraction-relaxation cycle is regulated by the interaction of many humoral regulating factors. Constrictors and dilators released from endothelial cells and neural and humoral factors contribute to the regulation of muscle tone. Changes in the nature of endothelial cells following damage or aging may also be responsible for the generation of vasospasm. More detailed future investigations are required to clarify the pathophysiological nature of vasospasm.

Acknowledgements. The authors wish to thank Dr. R.J. Timms, Birmingham University, UK. This work was supported by a Grants in Aid for Scientific Research from the Ministry of Health and Welfare, Japan.

References

1. Ebashi S, Koga R (1988) Desensitization of smooth muscle natural actomyosin. Proc Japan Acad 64B: 98–101
2. Ebashi S, Mikawa T, Kuwayama H, Suzuki H, Ikemoto H, Ishizaki Y, Koga R (1987) In: Sigman M, Somlyo AP, Stierens NN (eds) Ca^{2+} regulation of smooth muscle: Dissociation of myosin light chain kinase activity from activations of actin-myosin-interaction. Alan R Liss, New York pp 109–117
3. Ebashi S, Nonomiya Y, Nakamura S, Nakasone H, Kohama K (1982) Regulatory

---

Fig. 10. Effects of nicorandil on: **A** membrane potential changes measured from smooth muscle cell membranes of the porcine coronary artery; **B** and **C** on Ca-dependent K current measured from the rat portal vein using the patch clamp configuration (outside-out patch). **A** Effects of nicorandil on membrane potential measured in various concentrations of K (*control, closed circles,* and *open circles,* after application of nicorandil). **Ba** Effects of nicorandil on the 130 pS and 10 pS Ca-dependent K currents; **Bb** Effects of nicorandil on the 10 pS K current; **C** Traces of the 10 pS K channel current observed **a** before and **b** after applications of nicorandil (A from [28] and **B**, provided by S, Kajioka and K. Kitamura, unpublished observations)

mechanism in smooth muscle: Actin-linked regulation. Fed Proc 41: 2863–2867
4. Mikawa T, Y. Nonomura, Ebashi S (1977) Does phosphorylation of myosin light chain have direct relation to regulation in smooth muscle? J Biochem 82: 1789–1791
5. Mikawa T, Nonomura Y, Hirata M, Ebashi S, Kakiuchi S (1978) Involvement of an acidic protein in regulation of smooth muscle contraction by the tropomyosin-leitotonin system. J Biochem 84: 1633–1636
6. Adelstein RS, Eisenberg E (1980) Regulation and kinetics of the actin-myosin-ATP interaction. Annu Rev Biochem 49: 921–956
7. Hartshorne DJ, Gorecka J (1980) Biochemistry of the contractile proteins of smooth muscle. In Bethesda MD (ed) Handbook of physiology the cardiovascular system Am Physiol Soc sect 2, vol II, pp 93–120
8. Ikebe M, Inagaki M, Naka M, Hidaka H (1988) Correlation of conformation and phosphorylation and dephosphorylation of smooth muscle myosin. J Biol Chem 263: 10698–10704
9. Saida K, Nonomura Y (1978) Characteristics of Ca^{2+}- and Mg^{2+}-induced tension development in chemically skinned smooth muscle fibers. J Gen Physiol 72: 1–14
10. Somlyo AV, Somlyo AP (1968) Electro mechanical and pharmacomechanical coupling in vascular smooth muscle. J Pharmacol Exp Ther 159: 129–159
11. Walsh MP (1985) Calcium regulation of smooth muscle contraction. In: Marine D (ed) Calcium and cell physiology. Springer Berlin Heidelberg New York Tokyo
12. Onishi H, Wakabayashi T (1982) Electron microscopic studies of myosin molecules from chicken gizzard muscle I: The formation of the intramolecular loop in the myosin tail. J Biochem 92: 871–879
13. Suzuki H, Onishi H, Takahashi K, Watanabe S (1978) Structure and function of chicken gizzard myosin. J Biochem 84: 1529–1542
14. Corinna B, Rüegg JC, Takai A (1988) Effects of okadaic acid on isometric tension and myosin phosphorylation of chemically skinned guinea pig taenia coli. J Physiol 398: 81–95
15. Gabella G (1981) Structure of smooth muscles. In : Bülbring E, Brading AF, Jones AW, Tomita T (eds) Smooth muscle: An assessment of current knowledge. Edward Arnold, London, pp 1–46
16. Burnstock G (1972) Purinergic nerves. Pharmacol Rev 24: 509–581
17. Burnstock G (1980) Cholinergic and purinergic regulation of blood vessels. In: Bethesda MD (ed) Handbook of physiology. The cardiovascular system. Am Physiol Soc sect 2, vol II, pp 567–612
18. Burnstock G (1981) Neurotransmitters and trophic factors in the autonomic nervous system. J Physiol 313: 1–35
18. Bevan JA, Brayden JE (1987) Nonadrenergic neural vasodilator mechanisms. Circ Res 60: 309–326
19. Burnstock G (1981) Purinergic receptor. Chapman and Hall, London
19. Bevan JA, Oriwo MA, Bevan RD (1986) Physiological variation in α-adrenoceptor mediated arterial sensitivity: Relation to agonist affinity. Science 234: 196–197
20. Burnstock G (1981) Development of smooth muscle and its innervation. In: Bülbring E, Brading AF, Jones AW, Tomita T (eds) Smooth muscle: An assessment of current knowledge. Edward Arnold, London, pp 431–458
21. Bolton TB, Large WA (1986) Are junction potentials essential? Dual mechanism of smooth muscle cell activation by transmitter released from autonomic nerves. Q J Exp Physiol 71: 1–28
22. Burnstock G, Costa M (1975) Adrenergic Neurones. Chapman, and Hall, London
23. Bülbring E (1955) Correlation between membrane potential, spike discharge and tension in smooth muscle. J Physiol 125: 302–315

24. Hamill OP, Marty A, Neher E, Sakmann B, Sigworth F (1981) Improved patch-clamp techniques for high-resolution current recording from cells and cell-free membrane patches. Pflügers Arch 391: 85–100

25. Neher E, Sakmann B (1976) Single-channel currents recorded from membrane of denervated frog muscle fibres. Nature 260: 799–802

26. Ito Y, Kuriyama H (1971) Membrane properties of the smooth-muscle fibres of the guinea pig portal vein. J Physiol 214: 427–441

27. Ito Y, Kitamura K, Kuriyama H (1979) Effects of acetylcholine and catecholamines on the smooth muscle cell of the porcine coronary artery J Physiol 294: 595–611

28. Itoh T, Furukawa K, Kajiwara M, Kitamura K, Suzuki H, Ito Y, Kuriyama H (1981) Effects of 2-nicotinamidoethylnitrate on smooth muscle cells and on adrenergic transmission in the guinea pig and porcine mesenteric arteries. J Pharmacol Exp Ther 218: 260–270

29. Kajiwara M (1982) General features of electrical and mechanical properties of smooth muscle cells in the guinea pig abdominal aorta. Pflügers Arch 393: 109–117

30. Kajiwara M, Kitamura K, Kuriyama H (1981) Neuromuscular transmission and smooth muscle membrane properties in the guinea pig ear artery. J Physiol 315: 283–302

31. Komori K, Chen G, Suzuki H (1989) Mechanisms of inhibitory noradrenergic transmission in the rabbit facial vein. Pflügers Arch 413: 359–364

32. Komori K, Lorenz RR, Vanhoutte PM (1988) Nitric oxide, ACh and electrical and mechanical properties of canine arterial smooth muscle. Am J Physiol 255: H207–H212

33. Komori K, Suzuki H (1987) Heterogeneous distribution of muscarinic receptors in the rabbit saphenous artery. Br J Pharmacol 92: 657–664

34. Casteels R (1981) Membrane potential in smooth muscle cells. In: Bulbring E, Brading AF, Jones AW, Tomita T (eds) Smooth muscle: An assessment of current knowledge. Edward Arnold, London, pp 105–126

35. Creed KE (1979) Functional diversity of smooth muscle. Br Med Bull 35: 243–247

35. Caffrey JM, Josephson IR, Brown AM (1986) Calcium channels of amphibian and mammalian aorta smooth muscle cells. Biophys J 49 1237–1242

36. Kuriyama H, Ito Y, Suzuki H, Kitamura K, Itoh T (1982) Factors modifying contraction-relaxation cycle in vascular smooth muscles. Am J Physiol 243: H641–H662

37. Benham CD, Bolton TB (1986) Spontaneous transient outward current in single visceral and vascular smooth muscle cells of the rabbit. J Physiol 381: 385–406

38. Benham CD, Bolton TB, Lang RJ, Takewaki T (1985) The mechanism of action of Ba^{2+} and TEA on single Ca^{2+}-activated K^+ channels in arterial and intestinal smooth muscle cell membrane. Pflugers Arch 403: 120–127

39. Benham CD, Bolton TB, Lang RJ, Takewaki T (1986) Calcium activated potassium channels in single smooth muscle cells of rabbit jejunum and guinea pig mesenteric artery. J Physiol 371: 45–67

40. Berger W, Grygorcyk R, Schwarz W (1984) Single K^+ - channel in smooth muscles in membrane evaginations of smooth muscle cells. Pflugers Arch 402: 18–23.

41. Inoue R, Kitamura K, Kuriyama H (1985) Two Ca-dependent K-channels classified by application of tetraethylammonium distribute to smooth muscle membranes of the rabbit portal vein. Pflugers Arch 405: 173–179

42. Inoue R, Okabe K, Kitamura K, Kuriyama H (1986) A newly identified Ca^{2+} dependent K^+ channel in the smooth muscle membrane of single cells dispersed from the rabbit portal vein. Pflugers Arch 406: 138–143

43. Ohya Y, Kitamura K, Kuriyama H (1987) Modulation of ionic currents in smooth

muscle balls of the intestine by intercellularly perfused ATP and cyclic AMP. Pflugers Arch 408: 465–473

44. Ohya Y, Terada K, Kitamura K., Kuriyama H (1986) Membrane currents recorded from a fragment of rabbit intestinal smooth muscle cells. Am J Physiol 251: C335–C346

45. Okabe K, Kitamura K, Kuriyama H (1987) Features of 4-aminopyridine sensitive outward current observed in single smooth muscle cells from the rabbit pulmonary artery. Pflugers Arch 409: 561–568

46. Sadoshima S, Akaike N, Tomoike H, Kanaide H, Nakamura M (1988) Ca-activated K channel in cultured smooth muscle cells of rat aortic media. Am J Physiol 225: H410–H418

47. Brayden JE, Large WA (1986) Electrophysiological analysis of neurogenic vasodilation in isolated lingual artery of the rabbit Br J Pharmacol 89: 163–171

48. Johnsson B, Somlyo AP (1980) Electrophysiology and excitation contraction coupling. In: Bohr F, Somlyo AP, Sparks HD Jr (eds) Handbook of Physiology sect 2, vol II: The cardiovascular system Am Physiol Soc, Bethesda, pp 301–324

49. Standen NB, Quayle JM, Davis NW, Brayden JE, Huang Y, Nelson MT (1989) Hyperpolarizing vasodilators activate ATP-sensitive K^+-channels in arterial smooth muscle. Science 245: 177–180

50. Kajioka S, Oike M, Kitamura K (1990) Nicorandil opens a Calcium-dependent potassium channel in smooth muscle cells of the rat portal vein. J Pharmacol Exp Ther 254: 905–913

51. Bolton TB, Lim SP (1989) Properties of calcium stores and transient outward currents in single smooth muscle cells of rabbit intestine. J Physiol 409: 385–401

52. Hume JR, Lebranc N (1989) Macroscopic K currents in single smooth muscle cells of the rabbit portal vein. J Physiol 413: 49–73

53. Ohya Y, Terada K, Yamaguchi K, Inoue R, Okabe K, Kitamura K, Hirata M, Kuriyama H (1988) Effects of inositol phosphates on the membrane activity of smooth muscle cells of the rabbit portal vein. Pflugers Arch 412: 382–389

54. Sakai T, Terada K, Kitamura K, Kuriyama H (1988) Ryanodine inhibits the Ca-dependent KL current after depletion of Ca stored in smooth muscle cells of the rabbit ileal longitudinal muscle. Br J Pharmacol 95: 1089–1100

55. Ohya Y, Terada K, Kitamura K, Kuriyama H (1987) D600 blocks the Ca^{2+} channel from the outer surface of smooth muscle cell membrane of the rabbit intestinal and portal vein. Pflügers Arch 408: 80–82

56. Akaike N, Kanaide H, Kuga T, Nakamura M, Sadoshima J, Tomoike H (1989) Low-voltage-activated calcium current in rat aorta smooth muscle cells in primary culture. J Physiol 416: 141–160

57. Bean BP, Sturek M, Puga A, Hermsmeyer K (1986) Calcium channels in muscle cells isolated from rat mesenteric arteries: Modulation by dihydropyridine drugs. Circ Res 59: 229–235

58. Friedman ME, Suarez–Kurts G, Kaczorowski GJ, Katz GM, Reuben JP (1986) Two calcium currents in a smooth muscle cell line. Am J Physiol 250: H699–703

59. Loirand G, Pacaud P, Mironneau C, Mironneau J (1986) Evidence for two distinct calcium channels in rat vascular smooth muscle cells in short–term primary culture. Pflugers Arch 407: 566–568

60. Nowycky MC, Fox AP, Tsien RW (1985) Three types of neuronal calcium channel with different calcium agonist sensitivity. Nature 316: 440–443

61. Kawashima Y, Ochi R (1987) Two types of calcium channels in isolated vascular smooth muscles. Jpn J Physiol 49: 369

62. Worley III JF, Deitmer JW, Nelson MT (1986) Single nisoldipine-sensitive calcium channels in smooth muscle cells isolated from rabbit mesenteric artery. Proc Natl Acad Sci USA 83: 5746–5750

63. Furukawa K, Itoh T, Kajiwara M, Kitamura K, Suzuki H, Ito Y, Kuriyama H (1981) Vasodilating actions of 2-nicotin amidoethyl nitrate on porcine and guinea-pig coronary artery. J Pharmacol Exp Ther 218: 248–259

63. Benham CD, Hess P, Tsien RW (1987) Two types of calcium channels in single smooth muscle cells from rabbit ear artery studied with whole-cell and single-channel recordings. Circ Res 61 (Suppl 1): 10–16.

64. Yatani A, Seidel CL, Allen J, Brown AM (1987) Whole-cell and single-channel calcium currents of isolated smooth muscle cells from saphenous vein. Circ Res 60: 523–533

65. Yoshino M, Someya T, Nishino A, Yabu H (1988) Whole-cell and unitary Ca channel currents in mammalian intestinal smooth muscle cells: Evidence for existence of two types of Ca channels. Pflugers Arch 411: 229–231

66. Holman ME (1970) Junction potentials in smooth muscle. In: Bülbring E, Brading AF, Jones AW, Tomita T (eds) Smooth muscle. Edward Arnold, London, pp 244–288

67. Holman ME, Surprenant AM (1979) Some properties of the excitatory junction potentials recorded from saphenous artery of rabbits. J Physiol 287: 337–351

68. Holman ME, Surprenant AM (1980) An electrophysiological analysis of the effects of noradrenaline and α-receptor antagonists on neuromuscular transmission in mammalian muscular arteries. Br J Pharmacol 71: 651–661

69. Kuriyama H, Makita Y (1984) The presynaptic regulation of noradrenaline release differs in mesenteric arteries of the rabbit and guinea pig. J Physiol 351: 379–396

70. Kou K, Ibengwe JK, Suzuki H (1984) Effects of alpha-adrenoceptor antagonists on electrical and mechanical responses of the isolated dog mesenteric vein to perivascular nerve stimulation and exogenous noradrenaline. Naunyn Schmiedebergs Arch Pharmacol 326: 357–363

71. Suzuki H (1983) An electrophysiological study of excitatory neuromuscular transmission in the guinea-pig main pulmonary artery. J Physiol 336: 47–59

72. Ganitkevich VYA, Shuba MF, Smironou SV (1986) Potential-dependent calcium inward current in a single isolated smooth muscle cell of the guinea-pig taenia caeci. J PHysiol 380: 1–16

73. Inoue Y, Oike M, Nakao K, Kitamura K, Kuriyama H (1990) Endothelin augments unitary calcium channel currents on the smooth muscle cell membrane of guinea-pig portal vein. J Physiol 423: 171–191

74. Bolton TB (1979) Mechanisms of action of transmitters and other substances on smooth muscle. Physiol Rev 59: 606–718

75. Benham CD, Tsien RW (1987) A novel receptor-operated Ca^{2+}-permeable channel activated by ATP in smooth muscle. Nature 328: 275–278

76. Droogmans G, Declerck I, Casteels R (1987) Effects of adrenergic agonists on Ca^{2+}-channel currents in single vascular smooth muscle cells. Pflugers Arch 409: 7–12

77. Nelson MT, Standen NB, Brayden JE, Worley III JF (1988) Noradrenaline contracts arteries by activating voltage-dependent calcium channel. Nature 336: 382–385

78. Okabe K, Kitamura K, Kuriyama H (1988) The existence of a highly tetrodotoxin sensitive Na channel in freshly dispersed smooth muscle cells of the rabbit main pulmonary artery. Pflugers Arch 411: 423–428

79. Amédée T, Renaud JF, Jmari J, Lombert A, Mironneau J (1986) The presence of Na$^+$ channel in myometrial smooth muscle cells is revealed by specific neurotoxins. Biochem Biophys Res Commun 137: 675–681

80. Sturek M, Hermsmeyer K (1986) Calcium and sodium channels in spontaneously contracting vascular muscle cells. Science 233: 475–478

81. Bell C (1968) Dual vasoconstrictor and vasodilator innervation of the uterine arterial supply in the guinea pig. Circ Res 23: 279–289

82. Bell C (1969) Transmission from vasoconstrictor and vasodilator nerve to single smooth muscle cells of the guinea pig uterine artery. J physiol 205: 695–708

83. Nakazato Y, Ohya Y, Sigei T, Uematsu T (1982) Extrinsic innervation of the canine abdominal vena cava and the origin of cholinergic vasoconstrictor nerve. J Physiol 328: 191–203

84. Kuriyama H, Suzuki H (1981) Adrenergic transmissions in the guinea-pig mesenteric artery and their cholinergic modulations. J Physiol 317: 383–396

85. Cheung DW (1982) Two components in cellular response of rat tail arteries to nerve stimulation. J Physiol 328: 461–468

86. Hirst GDS, Edwards FR (1989) Sympathetic neuroeffector transmission in arteries and arterioles. Physiol Rev 69: 546–604

87. Hirst GDS, Neild TO (1980) Evidence for two populations of excitatory receptors for noradrenaline on arteriolar smooth muscle. Nature 283: 767–768

88. Hirst GDS, Neild To (1981) Localization of specialized noradrenaline receptors at neuromuscular junctions on arterioles of the guinea pig. J Physiol 313: 343–350

89. Hirst GDS, Neild TO, Silverberg GD (1982) Noradrenaline receptors on the rat basilar artery. J Physiol 328: 351–360

90. Itoh T, Kuriyama H, Suzuki H (1981) Excitation-contraction coupling in smooth muscle cells of the guinea pig mesenteric artery. J Physiol 321: 515–535

91. Kuriyama H, Makita Y (1982) Modulation of neuromuscular transmission by endogenous and exogenous prostaglandins in the guinea pig mesenteric artery. J Physiol 327: 431–448

92. Langer SZ (1974) Presynaptic regulation of catecholamine release. Biochem Pharmacol 23: 1793–1800

93. Suzuki H, Mishima S, Miyahara H (1984) Effects of reserpine treatment on electrical responses evoked by perivascular nerve stimulation in the rabbit ear artery. Biomed Res 5: 259–266

94. Burnstock G, Kennedy C (1986) A dual function for adenosine 5'-triphosphate in the regulation of vascular tone. Circ Res 58: 319–350

95. Burnstock G, Kennedy C (1986) Purinergic receptors in the cardiovascular system. Prog Pharmacol vol 6 (2): 111–132

96. Byrne NG, Large WA (1986) The effects of α,β-methylene ATP on the depolarization evoked by noradrenaline ε-adrenoceptor response and ATP in the immature rat basilar artery. Br J Pharmacol 88: 6–8

97. Cheung DW, Fujioka M (1987) Inhibition of the junction potential in the guinea-pig saphenous artery by ANAPP$_3$. Br J Pharmacol 89: 3–5

98. Ishikawa S (1985) Actions of ATP and α, β, -methylene ATP in neuromuscular transmission and smooth muscle membrane of the rabbit and guinea pig mesenteric arteries. Br J Pharmacol 86: 777–787

99. Komori K, Suzuki H (1987) Electrical responses of smooth muscle cells during cholinergic vasodilation in the rabbit saphenous artery. Circ Res 61: 586–593

100. Miyahara H, Suzuki H (1987) Pre-and post-junctional effects of adenosine triphosphate on noradrenergic transmission in the rabbit ear artery. J Physiol 389: 423–440

101. Sneddon P, Burnstock G (1985) ATP as a co-transmitter in rat tail artery. Eur J Pharmacol 106: 149–152
102. Suzuki H (1985) Electrical responses of smooth muscle cells of the rabbit ear artery to adenosine triphosphate. J Physiol 359: 401–415
103. Nakao K, Okabe K, Kitamura K, Kuriyama H (1988) Characteristics of cromakalim-induced relaxation in the smooth muscle cells of guinea-pig mesenteric artery and vein. Br J Pharmacol 95: 795–804
104. Suzuki H (1989) Electrical activities of vascular smooth muscles in response to acetylcholine. Asia Pacific J Pharmacol 4: 141–150
105. Langer SZ (1981) Presynaptic regulation of the release of catecholamines. Pharmacol Rev 32: 337–362
106. Langer SZ, Lehmann J (1988) Presynaptic receptors on catecholamine neurones. In: Tren deleuburg U, Weiner N (eds) Handbook of experimental pharmacology. Catecholamines I vol 85 (1) Springer Berlin pp 419–507
107. Starke K (1987) Presynaptic α-autoreceptors. Rev Physiol Biochem Pharmacol 107: 73–146
108. Starke K, Göthert M, Kilbinger H (1989) Modulation of neurotransmitter release by presynaptic autoreceptors. Physiol Rev 69: 864–989
109. Starke K, Langer SZ (1979) A note on terminology for presynaptic receptors. In: Langer SZ, Starke K, Dubocorich ML (eds) Presynaptic receptors Pergamon, Oxford pp 1–3
110. Ishikawa T, Yanagisawa M, Kimura S, Goto K, Masaki T (1988) Positive inotropic action of novel vasoconstrictor peptide endothelin on guinea pig atria. Am J Physiol 255: H970–973
111. Majewski H (1983) Modulation of noradrenaline release through activation of presynaptic β-adrenoceptors. J Auton Pharmacol 13: 47–60
112. Makita Y (1983) Effects of prostaglandin I_2 and carbo cyclic thromboxane A_2 on smooth muscle cells and neuromuscular transmission in the guinea-pig mesenteric artery. Br J Pharmacol 78: 517–527
113. Makita Y (1984) Effects of adrenoceptor agonists and antagonists on smooth muscle cells and neuromuscular transmission in the guinea-pig renal artery and vein. Br J Pharmacol 80: 671–679
114. Casteels R, Kitamura K, Kuriyama H, Suzuki H (1977) The membrane properties of the smooth muscle cells of the rabbit main pulmonary artery. J Physiol 271: 41–61
115. Casteels R, Kitamura K, Kuriyama H, Suzuki H (1977) Excitation-contraction coupling in the smooth muscle cells of the rabbit main pulmonary artery. J Physiol 271: 62–79
116. Droogmans G, Raeymaekers L, Casteels R (1977) Electro- and pharmacomechanical coupling in the smooth muscle cells of the rabbit ear artery. J Gen Physiol 70: 129–148
117. Itoh T, Ueno H, Kuriyama H (1985) Calcium-induced calcium release mechanism in vascular smooth muscles: Assessments based on contractions evoked in intact and saponin-treated skinned muscles. Experientia 41: 989–996
118. Suematsu E, Hirata M, Hashimoto T, Kuriyama H (1984) Inositol 1,4,5-trisphosphate releases Ca^{2+} from intracellular store sites in skinned single cells of porcine coronary artery. Biochem Biophys Res Commun 120: 481–485
119. Suematsu E, Hirata M, Sasaguri T, Hashimoto T, Kuriyama H (1985) Roles of Ca^{2+} on the inositol 1,4,5-trisphosphate-induced release of Ca^{2+} from saponin-permeabilized single cells of the porcine coronary artery. Comp Biochem Physiol 82A: 645–649

120. Hashimoto T, Hirata M, Itoh T, Kanmura Y, Kuriyama H (1986) Inositol 1,4, 5-trisphosphate activates pharmacomechanical coupling in smooth muscle of the rabbit mesenteric artery. J Physiol 370: 605–618
121. Somlyo AV, Bond M, Somlyo AP, Scarpa A (1985) Inositol trisphosphate-induced calcium release and contraction in vascular smooth muscle. Proc Natl Acad Sci USA 82: 5231–5235
122. Su C, Bevan JA, Ursillo RC (1964) Electrical quiescence of pulmonary artery smooth muscle during sympathetic stimulation. Circ Res 15: 20–27
123. Ohya Y, Kitamura K, Kuriyama H (1987) Cellular calcium regulates outward currents in rabbit intestinal smooth muscle cell. Am J Physiol 252: C401–C410
124. Walker JW, Somlyo AV, Goldman YE, Somlyo AP, Trentham DR (1987) Kinetics of smooth and skeletal muscle activation by laser pulse photolysis of caged inositol 1, 4, 5-trisphosphate. Nature 327: 249–252
125. Yamamoto H, van Breemen C (1985) Inositol-1,4,5-trisphosphate release calcium from skinned cultured smooth muscle cells. Biochem Biophys Res Commun 130 (1): 270–274
126. Abdel-Latif AA (1986) Calcium-mobilizing receptors, polyphospho-inositides, and the generation of second messengers. Pharmacol Rev 38: 227–272
127. Berridge MJ (1984) Inositol trisphosphate and diacylglycerol as second messengers. Biochem J 220: 345–360
128. Berridge MJ (1987) Inositol trisphosphate and diacylglycerol: Two interacting sceond messengers. Annu Rev Biochem 56: 159–193
129. Berridge MJ, Irvine RF (1984) Inositol trisphosphate, a novel sceond messenger in cellular signal transduction. Nature 312: 315–321
130. Berridge MJ, Irvine RF (1989) Inositol phosphates and cell signalling. Nature 341: 197–205
131. Nishizuka Y (1984) The role of protein kinase C in cell surface signal transduction and tumor promotion. Nature 308: 693–698.
132. Nishizuka Y (1986) Studies and prospectives of protein kinase C. Scinece 233: 305–312
133. Nishizuka Y (1988) The molecular heterogeneity of protein kinase C and its implications for cellular regulation. Nature 344: 661–665
134. Streb H, Irvine RF, Berridge MJ, Schulz I (1983) Release of Ca^{2+} from a nonmitochondrial intracellular store in pancreatic acinar cells by inositol-1,4,5-trisphosphate. Nature 306: 64–69
135. Furchgott RF, Zawadzki JV (1980) The obligatory role of endothelial cells in the relaxation of arterial smooth muscle by acetylcholine. Nature 288: 373–376
136. Furchgott RF (1984) The role of endothelium in the responses of vascular smooth muscle to drugs. Annu Rev Pharmacol Toxicol 24: 175–195
137. De Mey JG, Gläye M, Vanhoutte PM (1982) Endothelium-dependent inhibitory effects of acetylcholine, adenosine triphosphosphate, thrombin and arachidonic acid in the canine femoral artery. J Pharmacol Exp Ther 222: 166–173
138. Van Breemen C, Aaronson P, Loutzenhiser R (1979) Sodium-calcium interactions in mammalian smooth muscle. Pharmacol Rev 30: 164–208
139. Ignarro LJ (1989) Biological actions and properties of endothelium-derived nitric oxide formed and released from artery and vein. Circ Res 65: 1–21
140. Ignarro LJ, Ballot B, Wood KS (1984) Regulation of soluble guanylate cyclase activity by porphyrins and metalloporphyrins. J Biol Chem 259: 6201–6207
141. Ignarro LJ, Gruetter CA (1980) Requirement of thiols for activation of coronary arterial guanylate cyclase by glyceryl trinitrate and sodium nitrite: Possible involvement of S-nitrosothiols. Biochim Biophys Acta 631: 221–231

142. Ignarro LJ, Harbison RG, Wood KS, Kadowitz PJ (1986) Activation of purified Soluble guanylate cyclase by endothelium-derived relaxing factor from intrapulmonary artery and vein: stimulation by acetylcholine, bradykinin and arachidonic acid. J Pharmacol Exp Ther 237: 893–900

143. Ignarro LJ, Kadowitz PJ (1985) Pharmacological and physiological role of cyclic GMP in vascular smooth muscle relaxation. Ann Rev Pharmacol 25: 171–191

144. Ignarro Lj, Lippton H, Edwards JC, Baricos WH, Hyman AL, Kadowitz PJ, Gruetter CA (1981) Mechanism of vascular smooth muscle relaxation by organic nitrates, nitroprusside and nitric oxide: evidence for the involvement of S-nitrosothiols as active intermediates. J Pharmacol Exp Ther 218: 739–749.

145. Murad F (1986) Cyclic guanosine monophosphate as a mediator of vasodilation. J Clin Invest 78: 1–5

146. Rapoport RM, Murad F (1983) Agonist-induced endothelium-dependent relaxation in rat thoracic aorta may be mediated through cGMP. Circ Res 52: 352–357

147. Rapoport RM, Draznin MB, Murad F (1982) Sodium nitroprusside-induced protein phosphorylation in intact rat aorta is mimcked by 8-bromo cyclic GMP. Proc Natl Acad Sci USA 79: 6470–6474

148. Palmer PM, Ashton DS, Moncada S (1988) Vascular endothelial cells synthesize nitric oxide from L-arginine. Nature 333: 664–666

149. Palmer PM, Ferrige PG, Moncada S (1987) Nitric oxide release accounts for the biological activity of endothelium-derived relaxing factor. Nature 327: 524–526

150. Rees DD, Palmer RMJ, Hodson HF, Moncada S (1989) A specific inhibitor of nitric oxide formation from L-arginine attenuates endothelium-dependent relaxation. Br J Pharmacol 96: 418–424

151. Raeymaeker L, Hoffmann F, Casteels R (1988) Cyclic GMP-dependent protein kinase phosphorylates phospholambane in isolated sarcoplasmic reticulum from cardiac and smooth muscle. Biochem J 252: 269–273

152. Chen G, Suzuki H, Weston AH (1988) Acetylcholine releases endotheliumderived hyperpolarizing factor and EDRF from rat blood vessels. Br Pharmacol 95: 1165–1174

153. Nishiye E, Chen G, Kuriyama H (to be published) Regulation of vascular tone; endothelium derived regulating factors in the guinea pig coronary artery. Br J Pharmacol

154. Nishiye E, Nakao K, Itoh T, Kuriyama H (1989) Factors inducing endothelium-dependent relaxation in the guinea pig basilar artery as estimated from the action of hemoglobin. Br J Pharmacol 96: 645–655

155. Moncada S, Ferreira SH, Bunting S, Vane JR (1977) An enzyme isolated from arteries transformed prostaglandin endoperoxides to an unstable substance that inhibits platelet aggregation. Nature 263: 663–665

156. Yanagisawa M, Kurihara H, Kimura S, Tomobe Y, Kobayashi M, Mitsui Y, Yazaki Y, Goto K, Masaki T (1988) A novel potent vasoconstrictor peptide produced by vascular endothelial cells. Nature 332: 411–415

157. Shirahase H, Usui H, Kurahashi K, Fujiwara M, Fukui K (1987) Possible role of endothelial thromboxane A_2 in the resting tone and contractile responses to acetylcholine and arachidonic acid in canine cerebral arteries. J Cardiovasc Pharmacol 10: 517–522

158. Ishikawa S, Sperelakis N (1987) A novel class (H3) of histamine receptors on perivascular nerve terminals Nature 327: 158–160

159. Lippton H, Goff J, Hymanm A (1988) Effects of endothelin in the systemic and renal vascular beds in vivo. Eur J Pharmacol 155: 197–199

160. Warner T, De Nucci G, Vane JR (1988) Release of EDRF by endothelin in the rat isolated perfused mesentery. Br J Pharmacol 95: 723
161. Goto K, Kasuya Y, Matsuki N, Takuwa Y, Kurihara H, Ishikawa T, Kimura S, Yanagisawa M, Masaki T (1989) Endothelin activates the dihydropyridinesensitive, voltage-dependent Ca^{2+} channel in vascular smooth muscle. Proc Natl Acad Sci USA 86: 3915–3918
162. Ishikawa T, Yanagisawa M, Kimura S, Goto K, Masaki T (1989) Positive chronotropic effects of endothelin: A novel endothelium-derived vasoconstrictor peptide. Pflugers Arch 413: 108–110
163. Endo M (1977) Calcium release from the sarcoplasmic reticulum. Physiol Rev 57: 71–108
164. Itoh T, Izumi H, Kuriyama H (1982a) Mechanisms of relaxation induced by activation of β-adrenoceptors in smooth muscle cells of the guinea pig mesenteric artery. J Physiol 326: 475–493
165. Itoh T, Kajiwara M, Kitamura K, Kuriyama H (1982b) Roles of stored calcium on the mechanical response evoked in smooth muscle cells of the porcine coronary artery. J Physiol 322: 107–125
166. Itoh T, Kanmura Y, Kuriyama H (1985) A23187 increases calcium permeability of store sites more than of surface membranes in the rabbit mesenteric artery. J Physiol 359: 467–484
167. Itoh T, Kanmura Y, Kuriyama H, Sumimoto K (1986) A phorbol ester has dual actions on the mechanical response in the rabbit mesenteric and procine coronary arteries. J Physiol 375: 515–534
168. Kobayashi S, Kanaide H, Nakamura M (1985) Cytosolic free calcium transients in cultured vascular smooth muscle cells: Microfluorometric measurements. Science 229: 554–556
169. Iino M (1989) Calcium-induced calcium release mechanism in guinea pig taenia caeci. J Gen Physiol 94: 363–383
170. Iino M (1987) Calcium dependent inositol trisphosphate-induced calcium release in the guinea pig taenia caeci. Biochem Biophys Res Commun 142: 47–52
171. Takata Y, Kuriyama H (1980) ATP-induced hyperpolarization of smooth muscle cells of the guinea pig coronary artery. J Pharmacol Exp Therap 30: 708–728
172. Furuichi T, Yoshikawa S, Miyawaki A, Wada K, Maeda N, Mikoshiba K (1989) Primary structure and functional expression of the inositol 1,4,5-trisphosphate-binding protein p 400. Nature 342: 32–38
173. Kobayashi S, Somlyo AV, Somlyo AP (1988) Heparin inhibits the inositol, 1,4,5-trisphosphate-dependent, but not the independent, calcium release induced by guanine nucleotides in vascular smooth muscle. Biochem Biophys Res Commun 153: 625–631
174. Somlyo AP, Walker JW, Goldman YE, Trenthan DR, Kobayashi S, Kitazawa T, Somlyo AV (1988) Inositol trisphosphate, calcium and muscle contraction. Philos Trans R Soc Lond [Biol] 320: 399–414
175. Eggermont JA, Vrolix M, Wuytack F, Raeymaekers L, Casteels R (1988) The $(Ca^{2+}\text{-}Mg^{2+})$-ATPase of the plasma membrane and of the endoplasmic reticulum in smooth muscle cells and their regulation. J Cardiovasc Pharmacol 12: 551–555
176. Eggermont JA, Vrolix M, Raeymaekers L, Wuytack F, Casteels R (1988) Ca^{2+}-transport ATPase of vascular smooth muscle. Circ Res 62: 266–278
177. Tada M, Katz AM (1982) Phosphorylation of the sarcoplasmic reticulum and sarcolemma. Annu Rev Physiol 44: 401–423
178. Vrolix M, Raeymaekers L, Wuytack F, Hoffmann F, Casteels R (1988) Cyclic

GMP-dependent protein kinase stimulates the plasmalemmal Ca^{2+} pump of smooth muscle via phosphorylation of phosphatidylinositol. Biochem J 255: 855–863

179. Fujiwara T, Itoh T, Kubota Y, Kuriyama H (1988) Action of a phorbol ester on factor regulating contraction in rabbit mesenteric artery. Circ Res 63: 893–902

180. Chatterjee M, Murphy RA (1983) Calcium-dependent stress maintenance without myosin phosphorylation in skinned smooth muscle. Science 221: 464–466

181. Chatterjee M, Tejada M (1986) Phorbol ester-induced contraction in chemically skinned vascular smooth muscle. Am J Physiol 251: C892–C803

182. Murphy RA, Mras S (1983) Control of tone in vascular smooth muscle. Arch Intern Med 143: 1001–1006

183. Murphy RA, Aksoy MO, Dillon PF, Gerthoffer WT, Kamm KE (1983) The role of myosin light chain phosphorylation in regulation of the cross-bridge cycle. Fed Proc 42: 51–56

184. Kanmura Y, Itoh T, Kuriryama H (1987) Mechanisms of vasoconstriction induced by 9,11-epithio, 11,12-methano-thromboxane A_2 in the rabbit coronary artery. Circ Res 60: 402–409

185. Kitamura K, Kuriyama H (1979) Effects of acetylcholine on the smooth muscle cell of isolated main coronary artery of the guinea pig. J Physiol 293: 119–133

186. Suyama A, Kuriyama H (1984) Mechanisms of the ergonovine-induced vasoconstriction in the rabbit main coronary artery. Naunyn-Schmiedeberg's Arch Pharmacol 326: 357–363

187. Fujii K, Ishimatsu T, Kuriyama H (1986) Vasodilation induced by α-human atrial natriuretic polypeptide in rabbit and guinea pig renal arteries. J Physiol 377: 315–332

188. Sumimoto K, Domae M, Yamanka K, Nakao K, Hashimoto T, Kitamura K, Kuriyama, H. (1980). Actions of nicorandil on vascular smooth muscles. J Cardiovasc Pharmacol 10: S66–S77

189. Popescu LM, Panoiu C, Itinescu M, Nutu O (1985) The mechanism of cyclic GMP-induced relaxation in vascular smooth muscle cells. Eur J Pharmacol 107: 393–394

190. Stjärne L (1989) Basic mechanisms and local modulation of nerve impulse-induced secretion of neurotransmitters from individual sympathetic nerve varicosities. Rev Physiol Biochem Pharmacol 112: 1–137

191. Rüegg JC (1986) Calcium in muscle activation-A comparative approach. Springer, Berlin Heidelberg

192. Yanagisawa T, Kawada M, Taira N (1989) Nitroglycerin relaxes canine coronary arterial smooth muscle without reducing intracellular Ca^{2+} concentrations measured with fura-2. Br J Pharmacol 98: 469–482

193. Kamm KE, Stull JT (1985) The function of myosin and myosin light chain kinase phosphorylation in smooth muscle. Annu Rev Pharmacol Toxicol 25: 593–620

194. Bülbring E, Tomita T (1987) Catecholamine action on smooth muscle. Pharmacol Rev 39: 49–96.

195. Seki S, Suzuki H (1989) Comparison of the prejunctional α,β-adrenoceptor stimulating actions of adrenaline and isoprenaline in the dog mesenteric vein. Br J Pharmacol 97: 1324–1330

196. Kume H, Takai A, Tokuno H, Tomita T (1989) Regulation of Ca^{2+}-dependent K^+-channel activity in tracheal myocytes by phosphorylation. Nature 341: 152–154

197. Droogmans G, Calleweart G (1986) Ca^{2+}-channel current and its modification by the dihydropuridine agonist Bay K 8644 in isolated smooth musle cells. Pflugers Arch 406: 259–265

198. Makita Y, Kanmura Y, Itoh T, Suzuki H, Kuriyama H (1983) Effects of nifedipine derivatives on smooth muscle cells and neuromuscular transmission in the rabbit mesenteric artery. Naunyn Schmiedebergs Arch Pharmacol 324: 302–312

199. Terada K, Kitamura K, Kuriyama H (1987a) Blocking actions of Ca^{2+} antagonists on the Ca^{2+} channels in smooth muscle cell membrane of rabbit small intestine. Pflugers Arch 408: 552–557

200. Terada K, Nakao K, Okabe K, Kitamura K, Kuriyama H (1987) Action of the 1.4-dihydropyridine derivative, KW-3049, on the smooth muscle membrane of the rabbit mesenteric artery. Br J Pharmacol 92: 615–625

201. Terada K, Ohya Y, Kitamura K, Kuriyama H (1987) Actions of flunirizine, a Ca^{++} antagonist, on ionic currents in fragmented smooth muscle cells of the rabbit small intestine. J Pharmacol Exp Ther 240: 978–983

202. Okabe K, Terada K, Kitamura K, Kuriyama H (1987) Selective and long-lasting inhibitory actions of the dihydropyridine derivative, CV-4093, on calcium currents in smooth muscle cells of the rabbit pulmonary artery J. Pharmacol Exp Ther 243: 703–710

203. Hamilton TC, Weir SW, Weston AH (1986) Comparison of the effects of BRL 34915 and verapamil on electrical and mechanical activity in rat portal vein. Br J Pharmacol 88: 103–111

204. Weir SW, Weston AH (1986) Effects of apamin on response to BRL 34915, nicorandil, and other relaxants in the guinea pig taenia caeci. Br J Pharmacol 88: 113–120

205. Weir SW, Weston AH (1986) The effects of BRL 24915 and nicorandil on electrical and mechanical activity and on ^{86}Rb efflux in rat blood vessels. Br J Pharmacol 88: 121–128

206. Inoue T, Kanmura Y, Fujisawa K, Itoh T, Kuriyama H (1984) Effects of 2-nicotinamidoethyl nitrate (nicorandil; SG-75) and its derivatives on smooth muscle cells of the canine mesenteric artery. J Pharmacol Exp Ther 229: 793–802

207. Yamanaka K, Furukawa K, Kitamura K (1985) The different mechanisms of action of nicorandil and adenosine triphosphate on potassium channels of circular smooth muscle of the guinea pig small intestine. Naunyn Schmiedebergs Arch Pharmacol 331: 96–103

208. Gelband CH, Lodge NJ, Van Breemen CV (1989) A Ca^{2+}-activated K^{+} channel from rabbit aorta: modulation by cromakalim. Eur J Pharmacol 167: 201–210

Index